D0021574

HALFWAY TO ANYWHERE

THE TITLE OF this book, *Halfway to Anywhere,* was suggested by Tim Kyger, currently a staff member of the Senate Committee on Commerce, Science, and Transportation in Washington. It sums up where we are today with respect to real space transportation for everyone and comes from the statement—correct from the viewpoint of the energy and thus the rocket propellant required—I first heard in private, personal conversation with the late author, space advocate, and contemporary philosopher Robert A. Heinlein in 1950:

> *"Get to low-earth orbit and you're halfway to anywhere in the solar system."*

HALFWAY TO ANYWHERE

Achieving America's Destiny in Space

G. HARRY STINE

M. EVANS AND COMPANY, INC.

New York

M. Evans and Company, Inc.
216 East 49th Street
New York, New York 10017

Library of Congress Cataloging-in-Publication Data

Stine, G. Harry (George Harry), 1928-
    Halfway to anywhere : achieving America's destiny in space / G. Harry Stine. —
1st ed.
      p.   cm.
Includes bibliographical references and index.
ISBN 0-87131-805-9
    1. Aerospace planes—Research—United States—History
2. Aerospace planes—Political aspects—United States.   I. Title.
TL795.5.S7397 1996
629. 1' 2—dc20

96-18667
CIP

DESIGN BY BERNARD SCHLEIFER

Typeset by Not Just Another Pretty Face, NYC

Manufactured in the United States of America

First Edition

9 8 7 6 5 4 3 2 1

To
PHILIP BONO
1926 – 1992
who had the vision
but did not live to see it.

CONTENTS

FOREWORD

"Among the map makers of each generation are the risk takers,
those who see the opportunity,
seize the moment and expand man's vision into the future."
—Ralph Waldo Emerson

THE FUTURE IS RUNNING A LITTLE LATE. In 1964 and 1965 it was a foregone conclusion that we would be going to the moon on a steady basis and establishing a moon station there. We had also studied a voyage to Mars. But in 1967 President Johnson put the brakes on the entire program. We were literally all dressed up with nowhere to go. We had learned so much, expanded our awareness of the universe, traveled to the cosmic corridor, only to have the door slammed shut.

Although we continue to study space from near earth orbit, we haven't truly explored deep space since the days of the Apollo missions. We have become content to stay on this planet because we have become mired in economic issues, shortsighted vision, and party politics. Exploration has been curtailed because, quite frankly, it has become too expensive. However, as I review current affairs in technology and scientific exploration, I am optimistic. I have recently had the privilege of participating in the birth of the next generation of space travel, the Delta Clipper Single-Stage-to-Orbit reusable launch vehicle (RLV). This vehicle promises to make travel to space practical and affordable. Not only will companies be able to launch their communication satellites, but the RLV offers individuals the ability to send packages from New York to Paris in a mere 45 minutes, and medical materials from California to China in just 40 minutes. Imagine, for the price of a first-class ticket to Europe, we can all experience zero G! By establishing the spaceways as the new transportation infrastructure, space will become open to the public as a source of expanding economic opportunities, wealth, and new national and international purpose.

Communications have provided the impetus for the creation of a vehicle that will allow man to enrich his life. As we move forward toward the interweaving of cultures and the creation of global society, space will become the site of earth's great joint venture—a place where the concept of global peace may grow from a wishful dream to a concrete reality. Once again history invites us to take risks, to work as a team, and to explore man's full potential.

—CHARLES "PETE" CONRAD, JR.
Astronaut

FOREWORD

MANNED SPACE FLIGHT is thirty-five years old. Thirty-five years was also the time from the Wrights to the first jet. Aviation's development was spurred by unrelenting competition, military as well as commercial. By contrast, space access has been largely restricted to government bodies whose principal mission involved use of space, but was centered elsewhere, in military or information roles.

In consequence, the SL-7 booster that launched Gagarin is still in use, as is John Glenn's Atlas. Even the Shuttle, the first reusable spacecraft and really the space-age equivalent of the Wright Flyer, has been operating for fifteen years and is scheduled to last at least another fifteen.

The reliance on the tried and true has a downside. While all other travel is getting cheaper and more reliable, space flight is not. Sending a pound to orbit today costs $10,000, three or four times as much as it did twenty years ago, relative to other long haul trips, such as a flight to Asia. In consequence, the only thing worth bringing back from space today are bits and bytes, entertainment, communications, and observational data, not people or things.

Fortunately, the end of the cold war has opened a new door to space. Technologies that have been hidden are being released, and market opportunities that once were off limits are now available. The methods and the markets for a renaissance in space are consequently coming into focus. The management and the money to make it happen are beginning to appear.

Halfway to Anywhere is a chronicle of space entrepreneurship, a tale of a vision being brought into reality. The vision is affordable space, space made routine, space as a tool of business and as an available resource for the people rather than as an exotic activity for a few enormous government-regulated entities.

Entrepreneurship is difficult. *Halfway to Anywhere* reflects that painful reality. Businesses are built on profits, not technology. Profits come from sober and lucid evaluation of market-revenue potentials and offsetting cost assessments, combined with the willingness to take the financial and operational risks inherent in any high technology start up. Space travel is at the stage where that balance is becoming interesting. The transition to commercial space, to privately financed and operated space ventures is under way, beginning with the X-33 experimental spacecraft. That technology development program is a major milestone in the transition to the new world of reliable and economical scheduled access to space *Halfway to Anywhere* brings before our eyes.

—WOLFGANG H. DEMISCH
Managing Director
Institutional Equities
BT Securities Corporation

PREFACE

THIS BOOK REVEALS for the first time the revolution now under way that can, within a decade, make space transportation as economical, affordable, reliable, safe, and profitable as land, ocean, or air transportation. This revolution is taking place because of the intense, focused actions of a group of people who understood that no one was going to do what needed to be done unless they did it.

This has been a very strange book to write. The historical segment contains information never reported before. Some of it exists only in widely different places. Putting it together was a matter of research, scholarship, and digging through the files.

The last segment about the future of the Single-Stage-To-Orbit (SSTO) spaceship was fun because of my background as a technological forecaster, marketing research manger, and general soothsayer.

The most difficult part was the middle segment that concerns what's happening even as this is being written. Efforts to delay and even stop space access are unprecedented. The reader may have trouble believing it. However, I've been careful to document everything. Be aware that this middle, present-day segment necessarily stops with the situation as of December 31, 1995.

This book may resemble a personal memoir because, after fortyplus years of being involved in professional rocketry, model rocketry, scientific research, engineering and market consulting, television, technical writing, and a host of things that were interesting to me, I've never been one to sit on the sidelines of an interesting project that might improve What Is and help us get to What Can Be, including getting off this planet.

Back in 1957 when I was learning how to be a rocket engineer (not a rocket scientist), a friend at White Sands missile testing range told a reporter, "That boy just wants to get us out into space. He doesn't care what it's going to cost or how we do it—just so we do it!" Indeed, I was young then with many things to learn. One of them was that we must care about those factors. Because I didn't in the years before we did it at all, I might have contributed in some way to the Mess We're In with space transportation today. I apologize for screwing up. However, this time we've *got* to do space transportation right so it doesn't cost so much and everyone can go. We can.

If I do nothing more than document this incredible story of a turning point in our future, my own involvement will be justified to me. Maybe we won't make it. In which case, I can honestly say that I tried my best to make the future happen as I think it should. One of the messages of this book is that it's now a political and financial game, not a technical one.

Current events won't change the past and will affect the future only by delaying What Can Be. At worst, you may have to buy a ticket to space with eurodollars or yen because *someone* will provide economical, affordable, reliable, safe, profitable, on-demand space access.

This isn't a dream of a bunch of "space geeks" and aerospace engineers who want to do it because, like climbing Mount Everest, it's there to do. Instead of spending money on space activities, it's now a business challenge. An anonymous businessman once observed, "Never underestimate the American response to a business challenge." We got to where we are as Americans becuse we are a capitalistic frontier people. Don't forget what the frontier did for Americans in the last two centuries. As the late Gene Roddenberry said, space is the final frontier.

In the last chapter of his only nonfiction book, *Frontier*, the late Western author Louis L'Amour wrote, "I am often asked. 'Where is the frontier now?' The answer is obvious. Our frontier lies in outer space."

—G. Harry Stine
Phoenix, Arizona
January 1996

PART I

THE BIRTH OF
THE CONCEPT

THE CONCEPT OF the reusable Single-Stage-To-Orbit (SSTO) space-ship isn't new. Both Buck Rogers and Flash Gordon had them in the comic strips decades ago, and Hollywood special effects showed them on the silver screen as early as 1928. However, dreaming about something is different from making it happen. But if no one dreams about it, it never happens at all.

Part I presents the basics of the SSTO concept in non-technical terms—I won't use a single equation—and traces its development as a concept by many people. A surprising number of those who worked on these concepts are still active today and making SSTO spaceships become a reality.

Before 1988, however, the problems to be solved were techni-cal. For example, an SSTO built with 1960 structural, materials, and propulsion technologies would be able to carry only enough fuel to achieve a velocity of about 24,000 feet per second (16,400 miles per hour). This would permit it to fly from White Sands, New Mexico, to Thule, Greenland. In 1995, more than a third of a cen-tury later, we have composite materials of great strength and light weight, rocket engines of higher efficiency, improved reusable lightweight high temperature ceramic materials, and small, high-speed solid-state computers. These make it possible for us to build and operate a true SSTO capable of achieving an orbital velocity of

36,000 feet per second (24,500 miles per hour) with propellant reserves for entry and landing while carrying up to 10 tons of payload at 1/100th the cost of flying that payload on an expendable rocket as we've been doing since 1957. It's analogous to the difference between the Ford Tri-motor of 1930 with its ability to carry 13 passengers over a distance of 550 miles at 120 miles per hour and, 30 years later, the Boeing 707-320C jetliner that could take 189 passengers on an 8,000-mile trip at 600 miles per hour.

The unfolding of the saga also reveals that the reusable SSTO has followed the classic path from being technology-limited to politically-limited. This latter condition is usually caused by people either guarding the status quo or suffering from what might charitably be called a lack of testicular fortitude.

We'll see that the SSTO has gone through the various stages of acceptance experienced by any new idea. These were listed by Arthur C. Clarke in his book, *Profiles of the Future*. Clarke amended these in a personal letter to me dated February 26, 1994 as follows:

1. I never heard of such a crazy idea!
2. You may be right. So what?
3. I said it was a good idea all along!
4. I thought of it first!

With the SSTO, the time between Step 2 and Step 4 was surprisingly short. But it took more than 30 years to get past Step 1.

ONE

The Rooster Crows
at White Sands

AT 11:12:02 A.M. Mountain Standard Time on Saturday, September 11, 1993, a new kind of rocket lifted off at White Sands, New Mexico. It climbed to an altitude of 300 feet, hovered, moved horizontally 350 feet to the southwest, hovered again, lowered its landing gear, and descended to a soft and controlled landing 3.5 feet right and 1.5 feet short of its intended landing point on a plain concrete pad. The entire flight lasted 66 seconds.

So what? Big flight!

Yes, big flight.

It was the second flight of a sub-scale experimental version of a real spaceship deliberately designed as such and not as a modification of a long-range rocket-powered artillery shell.

History is full of other such portentous flights.

On December 17, 1903, at Kitty Hawk, North Carolina, Orville Wright flew a fragile wood-and-canvas aeroplane 120 feet in 12 seconds. On the second flight less than an hour later, Wilbur Wright flew the same machine a distance of 75 feet in 13 seconds. By noon, a fourth flight had been made to a distance of 852 feet in 59 seconds.

On March 16, 1926, on Aunt Effie's farm near Auburn, Massachusetts, Dr. Robert H. Goddard flew a liquid propellant rocket that took off, flew for three seconds, and crashed 184 feet from the launcher.

I hadn't been born when the flights were made at Kitty Hawk and Aunt Effie's farm. But I was at White Sands for the flight of the McDonnell Douglas Delta Clipper DC-X on September 11, 1993.

FIGURE 1-1: *Takeoff of the Delta Clipper DC-X on the first flight of a reusable rocket vehicle at Clipper Site, White Sands Missile Range, New Mexico, August 18, 1993. (Photo by White Sands Missile Range.)*

FIGURE 1-2: *The Delta Clipper DC-X hovering at an altitude of 150 feet during its first flight test. (Photo by White Sands Missile Range.)*

FIGURE 1-3: *The first landing of a reusable rocket vehicle. The Delta Clipper DC-X touches down. (Photo by White Sands Missile Range.)*

No one except the fifty-person flight test crew saw the first DC-X flight on August 18, 1993. About a thousand people saw the second one.

Why do I compare the flights of the DC-X to those of the Wright brothers and Doctor Robert H. Goddard?

Because all these flights smashed paradigms.

Very few people knew about the first airplane and liquid rocket flights at the time they occurred. Nothing about the Kitty Hawk flights appeared in the newspapers. Nothing about Goddard's first flight appeared in the newspapers. And very little about the flight of the DC-X showed up in the various news media. The first DC-X flight got a short segment on NBC Nightly News and CNN. The second flight got a short segment on CNN. But it wasn't big news.

However, the DC-X flights shook up people in the federal government and the aerospace industry in much the same way as the Soviet launch of Sputnik 1 on October 4, 1957.

The DC-X flight on September 11, 1993, was a watershed event. It showed we've been using a very expensive way to get to and from space for the last 35 years. It pointed clearly toward the future. And it showed once again that the way we make progress really hasn't changed.

Before Kitty Hawk, no one could fly.

Before Aunt Effie's farm, no one could get a liquid propellant rocket to fly.

Afterward, nearly everyone who wanted to do these things could and did. And performance improved dramatically.

Before the DC-X flights—please note the plural—a rocket was a single-flight piece of ammunition (except for model rockets). The second flight of the DC-X was probably more significant than the clandestine first flight (euphemistically called a "bunny hop" or "equivalent to an experimental aircraft's first high-speed taxi test" by the project spokespeople). The second flight proved it was possible to build a *reusable* rocket. It also proved beyond a doubt that the first flight wasn't a fluke.

Almost as a sideline, it showed that a major breakthrough in technology could be carried out in 21 months for only $60 million by a team of about 100 people. The DC-X is like Admiral Jackie

Fisher's H.M.S. *Dreadnought*, a British battleship built in 14 months whose launch in 1906 immediately made all other warships obsolete. The DC-X was believed to be impossible in an era of multi-billion dollar space projects, a lot of which create only mountains of paper, take ten years or more, and provide jobs for techno-bureaucrats. In fact, some anti-DC-X "experts" had to eat the word "impossible." But they never admitted to changing their minds.

The DC-X flight also spelled the beginning of the end of long-range artillery shells—modified intercontinental ballistic missiles (ICBMs)—used as space launch vehicles. This has caused some problems among those who favor these dinosaurs.

The DC-X showed that we could accomplish in the real world what authors had written about for decades: A space ship that takes off vertically under rocket power, maneuvers horizontally under rocket power, and lands vertically under rocket power.

Furthermore, the "thread of history" ran through the DC-X flights. The "flight manager" was Charles "Pete" Conrad who had already made one controlled rocket-powered landing and takeoff on the Moon.

The reason that the first flights of the DC-X are right in the center lane of technological progress stems from what such technological visionaries as the late Robert Heinlein, the late Herman Kahn, the late Dandridge M. Cole, Dr. Robert Prehoda, Dr. J. Peter Vajk, Arthur C. Clarke, and I had been saying for 40 years: "Technologists and the business people who hire them prefer to improve familiar technology by a fraction of a percent than to gamble on a major improvement. They manage to make marginal improvements on old technology just before it becomes obsolete."

Sometimes, it isn't new technology that does the trick. The novel application of *old technology* can do it, too. Or a combination of old and new technologies with a new philosophy that changes the entire paradigm or accepted way of doing things.

The DC-X is an excellent example of this.

I called it a "junkyard rocket" in a newspaper interview on the day of the second flight. This caused some expressions of disapproval on the part of the DC-X people who are understandably very proud of their rocket and what they accomplished. But it's true.

One of the rules of the DC-X program was, "Not one thin dime for research and development! Do it with equipment and technology that's off the shelf and exists!"

The Delta Clipper Team did just that. Fortunately, during the past 35 years of the national space program, a *lot* of technology has been developed. Some of it was used only once. Some of it wasn't useful at the time and was put back on the shelf. But it's there. It's paid for. It works.

Forced by the rules of the DC-X game, the Delta Clipper Team used the available and reliable Pratt & Whitney RL-10 rocket engines originally developed in 1960 for the Centaur rocket and used in the Saturn-I to power the S-IV stage.

The McDonnell Douglas F-15 fighter plane's navigation system and ring laser gyros and the McDonnell Douglas F/A-18 accelerometer and rate gyro package were appropriated and modified.

The McDonnell Douglas MD-11 airliner's autopilot and avionics were plugged into the system. Paul L. Klevatt, the DC-X program manager, remarked, "The DC-X really thinks it's an MD-11 operating on a strange flight plan."

Off-the-shelf Honeywell equipment such as the Global Positioning System (GPS) receiver and radar altimeter were used.

In the forty-foot Flight Operations Control Center—the "blockhouse" that's really the trailer from an 18-wheeler—were off-the-shelf commercial PC computers run by only three people. Modern software development tools such as ISI's MATRIX SystemBuild, AutoCode/AutoDoc, and AC-100 hardware-in-the-loop software prototyping tools were used to develop the flight control algorithms representing over 30,000 lines of Ada language code that makes the DC-X a "smart rocket." Furthermore, to shatter another paradigm, the software was delivered ahead of schedule and under budget, virtually unheard-of in the software industry.

The tankage was built by Chicago Bridge & Iron Services, Inc.

The composite graphite-epoxy aeroshell came from Scaled Composites, Burt Rutan's company at Mojave, California, that built the Voyager round-the-world airplane.

Some of the hinges and hatch-closing springs around the DC-X tail used hardware that came from Home Depot, Kmart, Tru-Value, and other ordinary sources. The watchword of the project managers was, "To hell with NASA qual and MIL specs! (Translation from techie shorthand: "National Aeronautics and Space Administration qualification and military specifications") Will it work? This is an experimental rocket, not an operational one! We have cost and time deadlines to meet!"

Where subcontractors didn't have or couldn't find off-the-shelf equipment and technology, McDonnell Douglas engineers scrounged through the aerospace junkyards on the West Coast. They looked for and found such items as titanium pressure spheres, hydraulic actuators, and other components designed and built for other aerospace vehicles.

The DC-X is indeed a "junkyard rocket."

The engineers and technicians also encountered all sorts of problems and troubles that plague engineers, technologists, and shade-tree mechanics who try to make equipment work the way it's supposed to. They ran into failures, glitches, and show-stoppers. To anyone who's worked on a car, a dishwasher, an airplane, or a rocket, for example, these occurrences are part and parcel of making anything work.

The second flight of the DC-X was another example of this.

Little of what went on behind the scenes was made public. But I had Air Force and McDonnell Douglas Delta Clipper Team people come up and say, "You may not believe this, but do you know what happened . . . ?"

The "Clipper Site" is on White Sands Space Harbor, a runway that has been used once for a space shuttle Orbiter landing. It can still be used, but NASA will do so only in a last-ditch emergency. The runway is merely a stretch of white gypsum sand. The wind blows a lot there. The white gypsum sand is carried by the wind into every nook and cranny of everything. NASA had to spend an inordinate amount of time cleaning white gypsum sand out of the Orbiter Columbia in 1982 after landing there. As a result, White Sands Space Harbor is now primarily used by shuttle astronauts for simulated orbiter landings flying the modified Gulfstream II jet aircraft. Before the shuttle came along, the place was known as the Northrup Strip that was often used to land the Northrop drone aircraft. (The misspelling of the name was carried over into the White Sands maps and has remained as the "Northrup Strip" ever since.)

When the wind blows at Northrup Strip, the sand blows. And both the wind and the sand blow a lot and at very high velocities. Especially during thunderstorms. During June, July, and August, thunderstorms occur almost daily. They are not the nice, soft thunderstorms often experienced elsewhere. These are "thunderboomers" of the most potent sort. They are violent. Lightning is bright blue and crackly. The rumble of the shuttle liftoff is nothing compared to the brisance and decibels of the resulting thunder. And the rain that comes out of these storms can only be described as "intense," dropping as much as three to four inches in less than an hour. On the level gypsum sand flats, this rainwater can't be described as "runoff" because it hasn't got any place to run to. So it puddles deeply in any low spot.

A few days before the critical September flight test, a thunderboomer visited the Clipper Site. It blew the doors off the canvas-covered "hangar" that's rolled over the DC-X between flights. The rainwater filled the trenches where the cryogenic feed lines run from the separate areas for the liquid oxygen (LOX) and liquid hydrogen (LH2) storage tanker trailers to the launch stand. The White Sands fire department had to come out to the Clipper Site and pump the water out of the ditches.

The DC-X spent four months at the NASA White Sands Test Facility for static testing the rocket engines and more than a month at the Clipper Site before it was flown for the first time. This didn't keep the Delta Clipper Team from encountering familiar (to me) old problems on September 11, 1993. I was fortunate to be next to someone who had a hand scanner tuned to the command frequency, so I heard it all.

By the time the crowd of about a thousand VIPs and media people got to the viewing site at 8:00 A.M. after a departure from Las Cruces on Grey Line buses at oh-dark-thirty, the pre-flight check was well under way. Liftoff was scheduled for 9:00 A.M. It didn't happen that way, of course. It didn't bother me; this was an experimental program, and I didn't care if the experimental spaceship flew on schedule or not. (In fifteen years or so, people are going to complain loudly about delayed departures.)

Pete Conrad and Paul Klevatt didn't run a regular countdown. It may have sounded that way to the other VIPs, but I was listening to the command net on the scanner. They used a check list. They jumped time. Then they held time. Then they picked up time with a huge jump in the "countdown." Certain things had to be done by a certain time, and only the last minute or so was there a "countdown" of the sort NASA is married to. It was like checking out a Boeing 747-400 for a nonstop flight from New York to Australia.

I didn't have a tape recorder running, so the following reportage is accurate to the best of my recollection. I tried not to add to what is already a series of happenings that might be right out of fiction. Few readers would believe the fiction.

A computer in the Flight Operations Control Center trailer right behind the viewing site suddenly got the hiccups. They pulled boards and ran diagnostics. They didn't think it was the software. It had to be one of the boards. So they called the McDonnell Douglas factory at Huntington Beach, California, who said they'd fly a new computer right from Long Beach to Northrup Strip on a McDonnell Douglas corporate jet. Then someone in the trailer did something to the computer. No one would admit to what was done but I suspect the sudden application of the sole of a shoe or the impact of a hammer or the equivalent used to discipline balky computers. The computer began to work perfectly. They watched it closely for an hour or so while they were solving other little problems but the glitch never repeated itself. Sometimes, you have to scare technology into working. It's like when your car suddenly begins to work perfectly once you take it in to be repaired.

They encountered telemetry signal drop-outs. Telemetry is the way the rocket reports its status by radio to the ground crew. In a smart rocket like the DC-X during an experimental test program, telemetry glitches are serious problems. Conrad, the two others on the launch crew, and the few project types in the trailer didn't know if the DC-X was behaving itself and doing what it was supposed to be doing . . . or not. They'd never had telemetry drop-outs before. Then someone woke up to the fact that they'd never had a thousand people wandering around outside the trailer before, either. It turned out that VIPs were walking and standing in front of the receiving antennas. The people were moved out of the way, and the telemetry behaved itself once again.

Remote unmanned loading of liquid oxygen and liquid hydrogen—the only two propellants used in the DC-X—began at 10:04 A.M. This went well until a valve in the DC-X stuck. Repeated attempts to get it to work were to no avail. So one of the seven-man ground crew was told to pick up a hammer, climb in one of the Military Police jeeps with an enlisted man, go down to the DC-X three miles away, and hit the valve a couple of licks. Trying to solve valve problems around lots of liquid oxygen and liquid hydrogen isn't a fun assignment. As the jeep roared away, Conrad called it back. Just like old times, the valve had suddenly decided to work properly.

"We're losing on-board helium pressure," came the report.

"Must be a leak. Can we override the pressurization procedure and pump it up shortly before engine start?"

"We'll try. But we may not have enough post-landing pressure."

"We'll go with it as it is." It turned out they had more than enough post-flight pressure. Even in a test program, you don't want to use up redundancy without good reason.

Then the check list got down to the recovery generator. Once the DC-X lands and makes itself safe, its batteries could be low and it needs electric power for further post-flight actions. The generator was parked next to the VIP bleachers on the road, hitched as a trailer behind a van. This was an ordinary electric generator run by a gasoline engine, a standard piece of Army equipment loaned to the DC-X program by White Sands. The generator was scheduled to be started and running before launch so that it could be spirited down the road to the landing site and plugged into the bird after the landing. The DC-X had lots of redundant equipment built into it, but the project people had spent their limited money on that and not on non-flying redundant ground equipment. That created a problem.

The generator wouldn't start.

Then someone remembered to put gasoline in it.

Once it started, the ground crew got no indication of electrical output.

"The White Sands people say it was working fine when they shut it down last night."

"Get them over here!"

Silence on the net. Then: "They pulled the main circuit breaker last night. They never told us the generator had one. The breaker is sort of hidden. We pushed the breaker back in, and it's working fine now!"

"We've *got* to modify this check list," Conrad decided.

It was clear sailing from that point on. A few minutes before takeoff, the DC-X performed an internal automated systems check and decided it was ready to fly. The DC-X is a smart rocket with lots of on-board automation. Conrad had the option of aborting at any point using his PC. (He's known by the Delta Clipper Team as "Mousekewitz" Conrad, the man with the fastest mouse in the West.)

At 11:11:59 A.M., the rocket commanded engine start. In the first three seconds of engine operation, the on-board software checked out the four RL-10 engines and determined everything was okay at 30% thrust level. It then commanded throttle-up to 80% of rated thrust, and the DC-X lifted off its launch stand.

At 300 feet altitude, the computer throttled the engines back to cause the DC-X to hover. Except for the bright orange flame caused by the rocket exhaust burning the paint off the launch stand flame bucket in the first second or so, the RL-10 engines produce no visible exhaust flame. Occasionally, a streak of orange would appear in the exhaust as an engine was throttled, usually by cutting back the flow of oxidizer before the fuel, thereby creating a visible hydrogen-rich exhaust.

We're all used to seeing a rocket take off and go up and out of sight. It's strange and unusual to see a rocket lift off, climb, stop, and hover.

Then the DC-X began to move *sideways* in flight, swiveling its gymballed rocket engines about two degrees to provide directional control.

The muted rumble of the rocket engines could be heard about fifteen seconds after liftoff. It was not as loud as a Boeing 737-300 taking off.

Once the on-board satellite navigation receiver said the DC-X was over the landing point, the engines were throttled and it began its tail-first descent. The four landing legs popped out on schedule. It touched down at about three feet per second amidst orange flame from the fuel-rich throttled engines. Once it got a weight-on-gear signal, it shut itself down and began its self-safing procedures. It landed 3.5 feet right and 1.5 feet short of the landing pad center.

The caravan consisting of a fire engine and vans carrying the seven ground-crew members and towing the running generator left at once for the landing site. The report came back that everything was fine. The paint had been scorched on the downwind side of the DC-X. But Paul Klevatt said it would be cleaned up easily with a little green liquid industrial cleaner and a scrub brush. The DC-X would fly again in a week or so, continuing its test program during which the Delta Clipper Team would "push the edge of the envelope" just like the X-1 and X-15 programs.

Everyone cheered after the engines shut down. In spite of the low humidity, there wasn't a dry eye in the crowd. This included several well-known science fiction authors. Chalk it up to the bright sunlight.

There was a sense of seeing something important happen out over those blinding white sands that morning.

Some people dismissed it as a "stunt with no real value." They weren't there. They haven't followed the development of the reusable SSTO. They have no sense of history. And their comments indicate they have no dream, only a desire to protect their ancient turf.

However, as Program Manager Dr. Bill Gaubatz is fond of quoting, "Some things have to be believed to be seen."

Some say it was our next important step toward space travel for everyone. I believe this is true. Furthermore, it's coming along right about on schedule.

Furthermore, the way it was done had a *good* and a *right* feeling to it. *This is the way we've always developed new airplanes!*

It was confirmation that we do indeed know something about making technological progress. It proved that we can really make technology work for us in peaceful, frontier-expanding ways now that we're in the postwar era of the 45-Year War (or 75-Year War, depending on when you want to start counting). We no longer have to beat the Soviets. We only have to beat ourselves, those among us who complain that we've passed our peak, that technology is out of hand, that we can no longer do anything for less than 10 billion dollars spent over 10 years.

We can. We did. We just have to keep doing it. We aren't off this planet yet, and we have just begun to use technology wisely to help us solve our problems.

But it looks like we'll continue to do it in the same seemingly halting and stupid way we always have, making lots of mistakes and learning from them. Hey, that's just the way it's done best, that's all. Recently, we've spent too many years in which we haven't been allowed to make any mistakes.

We're back on track, at least in the area of space transportation.

Ralph Waldo Emerson wrote, *"We think our civilization near its meridian, but we are yet only at the cock-crowing and the morning star."*

However, this didn't happen overnight.

Nor is the path to the exciting and promising future clear.

But the dream is indeed clear, and we have a way to make it come true.

It's now time to stop talking about it and do it.

That's what this book is all about.

We can do it.

We must do it.

We will do it.

TWO

Disintegrating
Totem Poles

SINCE 1960 OR SO, an idea has become entrenched in the minds of all but a few rocket engineers: "Space travel is so difficult, dangerous, and expensive that only the government can afford it."

This thinking was eagerly supported by people in NASA and the aerospace companies because it set aside a very large chunk of turf for their exclusive use. If other people could be convinced that space travel was difficult, dangerous, and exorbitantly expensive, who would bother to compete with this elite group?

It also served to justify large annual federal expenditures for space activities, eliminating the need to reveal the true nature of the space program as a huge nationalized jobs program. Ample justification exists for this charge.

Lyndon B. Johnson saw the space program as a way to secure political support by using it to industrialize the South. It also provided Johnson's interests and supporters with a lot of money. NASA facilities were built on tracts of land that no one thought were valuable—except the Johnson interests. Mosquito infested wetlands (swamps to local people) south of Houston and northeast of New Orleans in Mississippi suddenly became active space centers.

The paradigm and the turf it delineated were fiercely protected by politicians, congressional staff members, aerospace managers, and NASA bureaucrats. Attempts to break the paradigm by outsiders who knew that space transportation didn't have to be difficult, dangerous, and expensive were stopped in some very vicious ways. Engineers and space enthusiasts went along because the government space program was perceived to be the only game in town. However, there were some people who didn't agree.

For example, in the 1980s when Martin Marietta Corporation announced publicly that it intended to commercialize the Titan launch vehicle, Lt. Gen. James B. Abrahamson, then in charge of the NASA space shuttle program and later the first director of the Strategic Defense Initiative Office (SDIO), is known to have visited Martin Marietta executive Norman Augustine and flatly threaten to withdraw further support from Martin Marietta as a potential NASA contractor.

The brahmins played rough. These contentions are amply documented by material in my own files.

These high priests of space had built a series of totem poles called expendable launch vehicles. They were sacred. They cost lots of money that they could dispense among their people.

But these space launch vehicles—the only ones used thus far for space operations—were designed and built with old technology based on obsolete ways of thinking.

They were and are nothing more than ammunition.

Ammunition is something that military officers and sportsmen consider to be expendable.

Every rocket-propelled vehicle that has flown into orbit as of 1995 has been totally or partially expendable. It has thrown away parts of itself as it climbed into space. Only the NASA space shuttle recovers parts for future re-use: the winged Orbiter and the two Solid Booster Rocket (SRB) casings. Recovering and refurbishing these casings has turned out to be more expensive than throwing them away.

As space visionary Arthur C. Clarke observed more than a quarter of a century ago, operating expendable space transportation vehicles is equivalent to building the ocean liner H.M.S. *Queen Elizabeth II*, sailing it once across the ocean, and scuttling it upon arrival at its first port of call. The basic economics of such an operation mean it's extremely expensive to book passage or ship cargo on such a one-use vessel.

Every human transportation system is based on reusable vehicles. It's the only way to keep the cost within reason.

Then why did space transportation evolve over the decades using converted ammunition?

The roots of early space transportation—meaning from 1942 until today—are buried in history.

The Treaty of Versailles that ended World War I sowed the seeds of the initial national space programs.

This Treaty prohibited Germany from manufacturing or using any artillery piece having a bore of more than 170 millimeters (6.69 inches). The best range that can be reached by a shell fired from a gun of this size is about 17 miles, a shorter distance than an army can march in a day. Furthermore, Germany was prohibited from having an air force. Thus, the German army had no means for "deep interdiction" of the enemy's rear areas.

But the Treaty of Versailles said nothing about long-range rockets for the simple reason that the solid-propellent barrage rockets developed in the nineteenth century by people such as Sir William Congreve in England had been rendered obsolete by rifled cannon.

Thus, in 1929, Captain Dpl. Ing. Walter R. Dornberger, an artillery officer of the German army, was assigned the task of developing a long-range artillery rocket, something not banned by the Treaty. He found Dr. Wernher von Braun and other members of the German Rocket Society building and flying primitive rockets near Berlin on a shoestring budget. Dornberger formed a team around von Braun. At Peenemünde on the Baltic Sea from 1940 to 1945, German rocket engineers developed the first long-range liquid-propellant rocket, the Aggregat-4 or "V-2" (as the German propaganda ministry tagged it). Its basic design criteria came directly from Dornberger's experience as an artillery officer: Deliver a payload ten times that of the legendary Paris Gun of World War I to twice the range with twice the accuracy.

The Germans did it. The V-2 was used during World War II against England, Paris, Antwerp, and the Remagen Bridge. A total of 3,590 V-2 rockets are known to have been launched against military targets in Europe. It was indeed a self-propelled artillery shell.

FIGURE 2-1: *The German V-2, the long-range rocket-powered artillery shell that became the ancestor of all today's expendable launch vehicles. This is V-2 Round #2 on the launcher in the Army Launch Area at White Sands Proving Ground, New Mexico, on April 16, 1946. (Photo by White Sands Missile Range from G. Harry Stine archives.)*

But even the Germans knew the A.4/V-2 could carry payloads other than explosive warheads. In fact, they had plans to use their rocket for lofting scientific measuring instruments high into the atmosphere. The V-2 could boost such instruments to altitudes in excess of 100 miles if everything worked right. And on their drawing boards were improved rockets capable of lofting warheads across the Atlantic Ocean or payloads into orbit around the Earth.

And at the end of World War II, German V-2s served as the foundation for rocket development in both the United States and the Soviet Union.

Engineers of any kind prefer to start with known "art" and make incremental, step-by-step improvements. This approach lowers risks. Thus, the V-2 was the point of departure for the development of the Intercontinental Ballistic Missiles (ICBMs) fielded by both the United States and the Soviet Union in the 1957–1959 time period. These ICBMs were a direct development of the German V-2 long-range artillery shell.

By 1957, both the United States and the Soviet Union had prototype ICBMs in initial testing phases. The Soviets succeeded in flying theirs first by only a matter of months.

The Soviets under Sergei Pavlovitch Korolev, the "Soviet von Braun," then converted their R-7 ICBM into a space launch vehicle. The R-7 lofted the first unmanned earth satellite, Sputnik 1, on October 4, 1957. It was also the basic booster for the first manned orbital flight made by Major Yuri Alekseyevich Gagarin in the Vostok 1 space capsule on April 12, 1961.

The United States followed suit, converting the Redstone, Thor, Atlas, and Titan ballistic missiles (artillery shells) into manned and unmanned space launch vehicles. It was a "quick and dirty" reaction to the USSR space feats that were considered victories in the battle between democracy and communism. The United States government could not permit the USSR's space achievements to go unchallenged. Nor could the federal government abide the fact that a group of supposedly clumsy peasants had achieved a technical breakthrough apparently beyond the capabilities of the United States. The engineering approach had to be implemented: Use what you have, convert it to do what you want, and shoot them rockets!

When the decision was made to go to the Moon, von Braun's quick solution was to build a bigger V-2 called Saturn V. He didn't have time to do anything else. If rocket engineers had done it the "right way," we would have developed reusable spaceships to build one or more space stations, then gone to the Moon from the space station. In fact, the details of this procedure were given in the March 22, 1952, issue of *Collier's* magazine. However, because of the urgency of the Apollo program, we were not permitted to do it that way.

The Apollo Saturn launch vehicles were upgraded artillery shells having the German V-2 as a common ancestor. An artillery shell is not a good human transportation vehicle. In a pinch— which is where the United States was in 1957–1961—engineers can modify artillery shells to carry brave test pilots. But the operational paradigms of an artillery officer are not those of an airline executive.

An artillery shell is built to be fired only once. It's deliberately thrown away. It must work right the first time because once the trigger is pulled there is no room for a mistake. The U.S. Navy's 16-inch shells were designed and manufactured to be stored for as long as 25 years; when loaded into battleship guns, every one of them was expected to work perfectly when fired.

However, with the materials, rocket motors, and space technology of the 1950s, it wasn't possible to design or build a ballistic missile with a range of more than a few hundred miles unless it incorporated a principle known as "staging."

The basic rocket equations were worked out by a Russian schoolteacher, Konstantin Edouardovich Tsiolkovsky, in the first decade of the 20th century. In the United States, Dr. Robert H. Goddard developed similar equations in the 1920s. So did rocket and space enthusiasts in Great Britain, France, Italy, and Germany.

It's not necessary to show the rocket performance equations here. They can be found in any astronautics text book. Basically, they show that in order to achieve a high speed (velocity), a rocket must consist mostly of rocket propellants—a fuel to burn and an oxidizer in which to burn the fuel.

An important factor is the velocity of the rocket's exhaust. The faster the flow of gases in the rocket exhaust, the less propellant is required to achieve the same final rocket velocity. Rocket engineers call this "specific impulse," a term applied to engine performance in somewhat the same manner as miles per gallon applies to automobile performance.

A rocket motor and propellant combination with a high specific impulse is automatically considered by engineers to be technologically superior. This may not be the case, but it's the paradigm with which they've worked. Specific impulse is given in terms of "seconds," but it's actually pounds of rocket thrust force divided by the weight of propellants consumed per second to produce that thrust. With a higher specific impulse propellant/motor combination, a rocket will achieve a higher velocity with a given amount of propellant.

The German V-2 propulsion system using ethyl alcohol and liquid oxygen had a specific impulse of about 280 seconds. Modern liquid propellant rocket engines using liquid hydrogen and liquid oxygen achieve a specific impulse of 425 seconds or more.

To achieve a velocity equal to that of its exhaust gases, a rocket's "propellant fraction"—the weight of the propellants divided by the maximum takeoff weight—has to be 0.66. The German V-2, built with 1930 technology and using steel instead of the aluminum that was in critical supply, had a propellant fraction of 0.66. Later rockets in the 1940 time period had propellant fractions of about 0.8.

In order for a rocket to reach a velocity necessary to put it into orbit, it must have a propellant fraction of 0.9 or better.

Engineers could not achieve this in 1940 with the best materials and structural technology then available. Even today, some engineers believe that a propellant fraction of 0.9 isn't attainable.

However, rocket engineers came up with a solution by asking the question: Why carry along the useless weight of empty propellant tanks and other rocket parts after they're empty or no longer needed?

If a rocket carries a smaller rocket as its payload, the smaller rocket can attain some very high speeds.

This is the principle of "staging" or dropping rocket parts no longer needed. Early rocketeers called such multi-part rockets "step rockets" and later "staged rockets."

Because recovering and reusing the lower part or stage of the rocket was considered complex and added to the weight, engineers designed multi-staged rockets where the lower stages were thrown away.

A simple analogy will show why this is expensive. If you want to drive a small pickup truck coast-to-coast without stopping for gas, you can load the truck with about 1,000 pounds of gas in 34 5-gallon cans. The weight of the empty fuel cans will be about 350 pounds. So as you empty each can into the gas tank en route, you toss it away to lighten the load and increase the gas mileage. At the end of the trip, the truck weighs perhaps 60% of its starting weight.

But throwing away cans costs money. You have to buy more cans if you make a return trip.

If you also throw away the truck at the end of the transcontinental trek, your return trip will be more expensive still.

It turns out that for multi-staged rockets, fuel is a small percentage of the operating cost. The money is eaten up by the need to replace a major part of the vehicle if you want to make another flight with it, assuming it's reusable.

Modern ballistic missiles and space vehicles have one or more stages that are thrown away in flight. This is not only expensive but creates a safety hazard, requiring that space launch facilities be located on sea coasts or in sparsely inhabited regions.

If airliners dropped parts in flight, airports would have to be remotely located and flight paths would have to traverse regions where the ground impact of airplane parts wouldn't present a hazard to people or property.

The demands of economy, safety, and affordability are greatly reduced if airplanes don't drop the engines they need for takeoff but are unnecessary for cruising flight. The same holds true of rocket-propelled space vehicles.

Dr. Jerry E. Pournelle describes the current multi-staged space launch vehicles as "disintegrating totem poles." This is an apt metaphor.

The philosophies and operational methods of the artillery officer (which Dornberger was) were unconsciously carried over into the space transportation vehicles of the national space programs.

Because people rode atop these artillery shells, an additional element of safety and reliability was demanded.

This became increasingly expensive in spite of a growing track record of successes. The occasional expensive accident or failure produced additional requirements for increased safety and reliability. This increased both the costs and time needed to prepare for flight.

It was time for a change, and this was realized by NASA even as Armstrong and Aldrin were walking on the Sea of Tranquility in 1969, having ridden there atop the world's largest artillery shell.

THREE

All Things to All People—At a Price

AFTER THE APOLLO PROGRAM, a totally reusable spaceship was seen by some engineers as necessary for further space activities. The United States had spent $12 billion getting to the Moon. That was considered to be too expensive. Even in 1969 that was a lot of money, even in the national capital. Senator Everett Dirksen immortalized himself by remarking, "A billion here and a billion there sooner or later adds up to real money." Therefore, in honor of this gentleman's long and distinguished career, we shall hereafter use the term "dirksen" as the equivalent of a billion dollars of government funding (not money because there's a difference).

Cheaper access to space was clearly seen as necessary. But it was pursued under the then-existing paradigm that space transportation had to be a government monopoly because it was so difficult, dangerous, and expensive. Actually, space access was difficult, dangerous, and expensive *because* it was a government monopoly.

The NASA space shuttle was conceived in the late 1960s as a totally reusable vehicle capable of taking any kind of payload to orbit and back. It was to be all things to all space ventures. However, a totally reusable space shuttle was thought to be too costly. In the budget battles, political compromises overcame engineering capabilities. "Make it cheaper!" came the order from Capitol Hill. The appropriations that accompanied this order reflected its strength.

Although these budgetary compromises produced a space shuttle that was cheaper to design and build *at the time*, the long-range effect was to reduce its safety, reliability, and responsiveness as well as producing a continuing escalation of operating costs.

The space shuttle was originally pegged at a price of about $3,250,000 per flight to deliver a 65,000-pound payload to orbit. However, the growing expense of the system was compounded by the need to maintain a very large standing army to handle operations by the time the first shuttle flight was made on April 12, 1981. Today, this army numbers between 15,000 and 40,000 people. The count will vary, depending on to whom you talk. The number quoted here comes from Washington congressional staffers.

Shuttle costs were increased again after the loss of the Challenger and the seven people aboard on January 26, 1986. More stringent quality controls were instituted. Additional forms and manuals were developed to produce a traceable "paper trail" that covered every person who handled or worked on every test that was done on every part of a shuttle all the way from the original vendor's plant to the launch pad. Additional tests were required for such subsystems as the Space Shuttle Main Engines (SSME) that propel the Orbiter. Additional parts had to be kept in inventory at the Cape, and all these parts had to have their own paper trails. When a shuttle Orbiter lands after a flight, it is de-certified for future flights and must be completely rebuilt, retested, and recertified.

By 1994, the cost of launching a space shuttle had increased from 0.00325 dirksens to an estimated 1.25 dirksens. This number comes from a Congressional staff member. For safety reasons, the maximum shuttle payload was reduced from 65,000 pounds to about 35,000 pounds per flight.

Furthermore, because of all the tight safety controls and standards now required by NASA, it takes millions of dollars and years of time to qualify a payload for a shuttle flight.

The space entrepreneurs initially saw the space shuttle as cheap access to space. This dream disappeared after 1986.

In addition, NASA got very picky about what can be taken up in the shuttle, including personal effects of the crew members. I don't know how astronaut James Weatherby managed to get the ashes of

Star Trek creator Gene Roddenberry into space aboard the Columbia in October 1992 because mission specialist Dr. Jay Apt told me that on his missions the crew couldn't even take along a copy of a "Far Side" cartoon book for "recreational reading"—i.e., in the john.

The NASA space shuttle system also was touted as an "operational system" that would offer regularly scheduled flights to orbit and back every week. It has degenerated into an experimental vehicle dependent upon the local weather for launch and landing. If everything goes well, a space shuttle flight can be made every two months.

However, nothing can make the space shuttle system acceptably safe. A few years ago, I shared a panel discussion platform with the director of one of the NASA space centers. He bragged that NASA's latest improvements would increase the shuttle's reliability to 95%. That means 1 out of 20 flights will run into serious trouble. I asked if he would get on any airliner with 95% reliability because, with such a reliability figure, more than a hundred airliners would crash *every day* on departure from New York's JFK airport! I also asked if he would buy, much less ride in, any automobile having 95% reliability. The response was not friendly.

As the saying goes, this is no way to run a railroad. In fact, no airline can operate in such a way. Nor can a spaceline.

NASA officials will never admit that the space shuttle isn't an operational system. Then what is it? Answer: It's a collection of four experimental reusable orbiters that are wearing out long before they were expected to do so.

The space shuttle should have been developed and tested as an experimental vehicle. The resultant data should have been used by commercial industry to develop truly economical, reliable, and safe spaceships. This didn't happen. Politics and bureaucracy got in the way of engineering and economics.

The existing paradigm also got in the way. The space shuttle was designed as a government operated space vehicle that was never intended to make money or even be commercially available. The government owned it. The government operated it. It was historically equivalent to the initial attempts of the U.S. Post Office to fly the airmail in 1919. Within a year, the Post Office had lost 31 of its 40 pilots.

Then why was the space shuttle developed as it was? Answer: Because of the way that the U.S. space program had developed, NASA needed a project that would maintain the huge standing technical army it had organized during the Apollo manned lunar landing program. It had allies in the aerospace industries who had formed large standing armies of their own to support Apollo. Shuttle was the logical (to them) follow-on to Apollo.

The Apollo program was the most complex activity ever undertaken by the human race. It dwarfed even Operation Overlord, the Allied landing in Normandy in 1944. Military officers organized and managed that one, and they were called in to recruit the space army and manage the Apollo program, the final battle in the technological war declared by the Soviet Union when they launched Sputnik 1. Neil Armstrong and Buzz Aldrin landed on the Moon on July 21, 1969 and the United States won that technological war. At the end of every war, the United States has always demobilized the army needed to win. However, in the case of the manned lunar landing program, the army wasn't disbanded. It's still there more than 25 years later, bigger than ever and devouring most of the NASA budget.

The United States government had created something equivalent to a National Airplane. Such a thing could have happened in the early part of the 20th century, especially after World War I. By playing the game of "let's suppose," perhaps we can get a better perspective on the present-day space program.

Let's suppose that, in 1919, a strong president and a willing Congress had decided that aviation was indeed the future of America. In World War I, the airplane had shown that it could do many things. Therefore, a National Airplane Program was instituted. The objective was to build an airplane that could be used for any purpose. It could carry passengers, perhaps at speeds up to 150 miles per hour. It would carry cargo or even bombs in case of war. It could be used to patrol our shores and shoot down any approaching enemy airplanes. It could be used for crop dusting. It could be used for aeronautical research. In short, the National Airplane would have many uses. This was necessary in order to justify its high cost and long development time. The objective was to have a fleet of these airplanes available by the year 1940 (this being the earliest that conservative aeronautical engineers were willing to estimate such a plane could be available, given the state of the art in 1919).

The Soviet Union produced such a national airplane: The Antonov An-2, NATO code name "Colt." It's the world's ugliest airplane. Powered by a huge 1,000 horsepower radial, air-cooled reciprocating engine, it's a biplane capable of carrying 14 passengers (or paratroopers) or a payload of 2,850 pounds (usually potatoes). It can fly 525 miles at the blistering speed of 105 miles per hour. It can operate from plowed fields, pastures, runways, snow fields, glaciers, or sandy deserts. It's used for crop dusting. Reports from American pilots who have flown it reveal that the Soviet Union must have trained its Olympic weight-lifters by teaching them to fly the An-2; moving the controls requires the strength of Sampson. In order for the pilot to get the tail up for takeoff, all passengers must move to the front of the cabin. No seat belts are provided; passengers are supposed to hang on to the exposed interior fuselage structure. Because of the racket of the huge engine on the front of the cavernous fuselage, An-2 pilots are chronically hard of hearing and conversation among the passengers is impossible.

However, the An-2 was useful in parts of the world where it was the only airplane available under the communist system of bean counting and currency exchange. More than 5,000 were built in the Soviet Union, and the airplane continues to be built in Poland. It has been sold in countries ranging from Afghanistan to Vietnam, usually to government or military operators because the An-2 (like nearly all Soviet commercial aircraft) isn't economical to operate.

One needs only to look at other government supported technological projects to speculate on the results. Read the results of the British government's Brabizon Committee, established during World War II. It was supposed to plan in detail the specific requirements for the commercial aircraft that British firms would build after the war. Only one of those committee-planned airliner designs was a modest success: 445 of the Vickers Viscount turboprop airliners were built and sold. (Boeing has delivered more than 2,000 737s and 1,000 747s.)

On the other hand, look at what really happened when the federal government stepped back, did only the risky research, and let private enterprise design and build airplanes. In 1915, the government created the National Advisory Committee for Aeronautics (NACA), the predecessor of NASA, whose job was to conduct research, provide data to the fledgling American airplane industry that had not produced a single military airplane during World War I, and eliminate perceived high technical risk by building and flying experimental vehicles. Today, except for specific cargos that require specific military cargo airplanes like the C-5A Galaxy, the Department of Defense charters commercial airliners for Military Airlift Command (MAC) and has available the Civil Reserve Aircraft Fleet (CRAF) of nearly 2,000 ordinary commercial airliners that can be mustered for military duty overnight. Otherwise, the government charters the flights it needs from commercial airplane operators.

When you board a comfortable jet airliner for a business trip, think of what it might have been like if the United States had developed a National Airplane that formed the basis for a federal airline ("Flytrak"?). If you had the money, you might be able to buy a ticket for an exorbitant price 18 months in advance, provided you'd completed a physical examination and six weeks of passenger training. And if a National Airplane happened to be flying where you needed to go that day far in the future.

That's equivalent to what we American taxpayers bought when NASA developed and operated our national space transportation system, the space shuttle.

NASA tried to sell the space shuttle system to a commercial operator in the late 1980s. It found no takers. No wonder! No company could possibly make any money with the space shuttle system!

The basic error in the space shuttle systems thinking was the artillery shell paradigm held over from the initial space program and ballistic missile development.

Thus the belief became entrenched in everyone's mind: "Space travel is so difficult, dangerous, and expensive that only the government can afford it."

The reasons why space travel has been difficult, dangerous, and expensive should be clear now.

If the United States had not over-reacted to the Soviet space program and thus opted to engage in an expensive, emergency, expedited, and accelerated lunar landing program for which there was no time to do anything but build upon a foundation of space launchers that were really long-range artillery shells, we might have developed yet another approach that was already starting in 1961: Flying manned airplanes into space with airline style operations.

The North American X-15 rocket plane, a vehicle openly tagged as an experimental or "X-ship," might have evolved into an experimental orbital vehicle by 1965–1967. In fact, I presented such a program in my 1957 book, *Earth Satellites and the Race for Space Superiority.*

But that sort of an incremental aviation approach was overwhelmed by the Apollo program.

However, the X-15 did get into space. Several X-15 pilots such as Neil Armstrong and Joe Engle got their astronaut wings by flying the X-15 higher than 50 miles.

Because of the pressures of the Cold War, we took the wrong track in the development of commercial space transportation. Because the space program stemmed from a matter of national prestige and therefore national security, cost was a secondary factor and access to space was controlled by the federal government. However, if we'd reacted in a more typically American manner, we might not have taken a path that led to a dead end in space with expensive, unreliable, unsafe, uneconomical, and controlled space access.

Reliable, economical, safe, on-demand space access is the key to the future in space for everyone, not just for government purposes. The various agencies of the federal government such as the Department of Defense and the National Oceanic and Atmospheric Administration (NOAA) need space access to carry out their responsibilities. But private, commercial entities such as telecommunications companies also want access.

If a shortage of space access capability exists in the United States or if space access is too expensive, the private sector must and will go elsewhere for it. And they have. They've gone to Russia, Europe, and China to get rides to space for their satellites. Even with the privatization of expendable American launch rockets, the United States' share of the space launch business is now 30% of what it was 25 years ago according to such sources as the 1994 Commercial Space Transportation Study.

There is no shortage of ideas about how to use space for providing products and services of value to people. Business people now realize they can make money by using space.

But we can't get there from here!

It's growing apparent that commercial airliner-type operation is the key to making space transportation affordable, responsive, reliable, and safe in contrast to using expendable rockets and partially-reusable spaceships that are the equivalent of national airplanes.

But it took a long time for that paradigm to be accepted. And the early advocates of reusable SSTO spaceships were ignored just like many aviation pioneers.

FOUR

The Genesis of
an Idea

WHERE DID THE CONCEPT OF the Single-Stage-To-Orbit spaceship come from?

And why discuss the past when the future seems so important?

To answer the second question: We navigate constantly into an unknown future with no charts or maps to guide us. We can know the direction we're going only by knowing the direction we've taken thus far. In navigation, this is called "dead reckoning." It comes from the term "deduced reckoning" and has nothing to do with the possibility that if you don't reckon correctly, you're dead (although this is always a possibility).

In addition, knowing the past can make us aware of pitfalls to avoid when problems arise—and they always do. It helps us avoid dead-ends and not reinvent the square wheel.

Finally, the people who have pioneered the ideas and concepts of the SSTO largely have been ignored or forgotten by aerospace historians. Attention has been focused on those who developed the expendable long-range artillery shells that have been the primary space transportation vehicles thus far. Part of the purpose of this book is to give credit to the work of these SSTO pioneers, many of whom didn't live to see their dreams become reality and whose work may otherwise be lost to history.

The SSTO concept has its origins in the literature of science fiction. This is natural. Most space travel ideas were written about in fiction long before they became reality.

Jules Verne's *Columbiad* space capsule in his classic 1865 novel, *De la Terre à la Lune (From the Earth to the Moon)* was an "SSTL" (Single-Shot-To-Luna) vehicle. It used the huge bullet shot to the Moon from a monster cannon. Verne's mathematics were correct. However, he neglected to account for air resistance about which little was then known. He also overlooked human tolerance to extremely high accelerations, another area where data wasn't available until the 1950s when Col. John Paul Stapp began his famous tests on rocket sleds. Today we know that the *Columbiad* would never have left the muzzle of the cannon because of air compression within the bore ahead of the accelerating vehicle. Its human occupants would have been smashed to jelly by the extremely high acceleration. To achieve lower, human-tolerable accelerations, engineers had to develop an artillery shell with a longer virtual gun barrel. This was called the rocket. Even at that, it took a century of engineering work before people could be shot into space with the descendants of Verne's space cannon.

Fortunately for potential space passengers, the Jules Verne era is coming to an end.

The beautiful and functional spaceship *Luna* in the 1950 motion picture, *Destination Moon*, written by Robert A. Heinlein, was an SSTO and more. The *Luna* was recoverable and probably reusable. It was a single-stage rocket. Again, the mathematics were all correct, as they always were in any Heinlein work. However, with the known state of the art of rocket technology at that time, Heinlein had to postulate the use of a nuclear rocket engine for the lunar voyage because only a nuclear rocket engine had the necessary efficiency (specific impulse) for this application. A nuclear reactor is used to vaporize a "working fluid" such as liquid hydrogen that is then accelerated out of an ordinary rocket nozzle.

Nuclear-powered spaceships were later shown to be feasible. The Kiwi—so called because this static-test proof-of-principle nuclear rocket engine was never intended to fly—and the flight-intended Nerva underwent extensive testing in a government research and development program until the nuclear test ban treaties prohibited further work on the concept.

FIGURE 4-1: *The Jules Verne Columbiad, the fictional cannon-shell prototype for all of today's expendable space launch vehicles.*

Figure 4-2: The beautiful nuclear-powered single-stage spaceship Luna *from the 1950 motion picture* Destination Moon, *designed by Robert A. Heinlein and Chesley Bonestell. (Photo from Stine archives.)*

Nuclear-powered SSTO spaceships provided the background for novels by such authors as Cyril Kornbluth and Lee Correy in the 1950s. Correy's designs were so realistic that photographs of his models appeared on several magazine and book covers of the period.

Although these author-engineers assumed that propulsion, materials, and structural technologies would progress in such a way that single-stage spaceships would become possible, they neglected a very important element:

Who's going to write the checks to buy such spaceships? And what will it cost to operate them? Can they be operated profitably?

The economics of space transportation were almost totally neglected save for one man, Robert Cornog, of the Ramo-Wooldridge Corporation. His trailblazing paper, "The Economics of Rocket-Propelled Airplanes," appeared in the *Aeronautical Engineering Review* in September and October, 1956.

Cornog was perhaps the first to state what is now a given: "If a nonrecoverable vehicle is used to deliver a load to a destination, it will be found that the cost of the structural parts of the vehicle, which must be included in the total delivery cost, is the dominant factor in determining the total delivery cost."

Cornog also based his economic analyses on comparable costs in the airline, railroad, and ocean-going transportation industries. In his paper, he pointed out:

> In most parts of the world, it is cheaper to buy, maintain, and service an oxcart than it is an airplane. In spite of this fact, there are more airplanes than oxcarts in most civilized areas of the world. The key to this paradox is the relative speed of transportation afforded by each vehicle.

He went on to say that the gross revenues or income of a transportation system are, in almost every case, directly proportional to the amount of service rendered. Maximizing gross income can be accomplished in two ways: (a) operate the equipment as much as possible, and (b) operate at high speeds. These principles are known and followed by the airline industry. However, in space transportation, we have now rediscovered Cornog's basic truths.

Cornog's classic paper lay forgotten for almost 40 years. Cornog can rightly be considered to be the Tsiolkovski and Goddard of space transportation economics.

Because space transportation is more than the technology involved in building and operating spaceships, readers should be aware that some people have known of these principles for nearly a half century. However, aerospace engineers ignored them because the government was footing the bill out of what appeared to be a bottomless purse of tax revenues. Cornog was writing about profitable commercial rocket transport operations.

And in the years before Sputnik, space planners weren't wedded to the ballistic missile as a space transportation vehicle. Some were laying groundwork that was buried for more than 50 years.

In October 1945, the U.S. Navy Bureau of Aeronautics proposed the development of what was called a High Altitude Test Vehicle (HATV), a single-stage-to-orbit rocket powered by liquid oxygen and liquid hydrogen. The proposal originated at the Naval Air Experimental Station in Philadelphia from one of their civilian engineers, Robert A. Heinlein (1907–1988). It was turned down by the Navy because the Naval Research Laboratory had been given the responsibility for high altitude rocket research.

In a paper presented before the American Rocket Society in 1956, Dr. William O. Davis (1920–1974), then a colonel with the Air Force Office of Scientific Research, took a green field approach to space transportation. He wrote:

> Let us forget for the moment that we ever heard of the rocket and guided missile. Let us assume that we will start from scratch with the science that is available to us and a knowledge of our objectives and design a space flight system from here. The basic requirements for a manned space flight system can be determined from a knowledge of human limitations, human physiology, engineering, economics, and common sense. If we were to list these basic requirements in order of their priority, they would be as follows:
>
> 1. The spaceflight system should be so designed and operated that there is a high probability that the human

beings aboard will be able to return to earth safely, either in case of a normal flight or in the case of an emergency.

2. The space flight system should be so designed and operated that there is a high probability that all equipment will survive the flight in workable condition.

3. There should be a high probability for flight success inherent in the system.

4. The manned space flight system should be compatible with the physical, biological, and mental limitations of human beings. There should be low acceleration but no zero gravity. Tolerable temperatures should be maintained in the system at all times. Flight times should be short, not over several weeks in duration. The entry profile for the return to the earth should be reasonable with low accelerations, low heating, constant control, and ability to select the landing point.

5. The manned space flight system should perform in an economical manner.

At this time, we haven't achieved all of the Davis criteria, but the development of the SSTO promises to make them viable with the exception of the zero gravity requirement, at least for the important Earth-to-orbit-and-return phase.

All of this work was forgotten after October 4, 1957, when the Soviet Union startled the world (but not the rocket engineers) by launching Sputnik 1 with the Korolev R-7 (8-K-63) ballistic missile.

This immediately moved space flight out of the realm of science fiction. It became a national security issue. Many of us involved in space advocacy work at that time believed Sputnik ensured that all future space activities would occur in a natural sequence of forecastable events because of the availability of government money to reduce technical risks.

In retrospect, President Dwight D. Eisenhower was correct in his assessment of the true threat to the security of the United States posed by Soviet space achievements. A recently released (1995) historical summary of one of the U.S. military surveillance satellite programs, Project Corona, revealed that no "missile gap" existed between the United States and the Soviet Union. However, this could not be revealed without letting the Soviets know just how good the American intelligence equipment really was.

Soviet Premier Nikita S. Khrushchev was as surprised as anyone at the world's reaction to the launching of Sputnik 1. However, Khrushchev made the best of it from the propaganda standpoint. His big mistake was challenging the United States to a technological war. In retrospect, the United States government should have stood aside and allowed the Soviets to fail, as their lunar landing program did in 1969. I'm aware that this isn't a popular stance to take, but the American response to the Soviet space spectaculars resulted in a course of action that probably set back commercial space transportation 20 to 30 years.

In the United States at the time, we were well along in our consideration of commercial spaceships.

Serious space forecasters such as Dandridge MacFarlane Cole (1921–1965) at the Martin Company and General Electric looked deeply into the future of space transportation, industrialization, and habitation. However, far-seeing as Cole was during his brief life, he too was strapped with the existing technical state of the art of materials and propulsion. Therefore, although Cole's large spaceships were single-stage vehicles, all of them were nuclear propelled. These became impractical after the signing of the Nuclear Test Ban Treaty of 1963.

If anyone can be considered to be the "father" of the SSTO, it is Philip Bono (1926–1992).

An aeronautical engineer specializing in structures and propulsion, Bono was one of those rare individuals who managed to survive decades of aerospace engineering without losing his ability to suggest interesting new approaches. In 1947, he went to work for what was then North American Aviation analyzing captured German V-2 rockets. He then spent nine years at Boeing working on a winged, reusable Air Force space vehicle called Dyna-Soar.

The X-20 Dyna-Soar was a one-man rocket-boosted orbital glider that was, in some ways, the precursor to the NASA space shuttle. Dyna-Soar was canceled by Robert McNamara's Department of Defense because no one could define a "mission" for it.

Defense experts believed that the ballistic Mercury, Gemini, and Apollo style of spacecraft were lighter, cheaper, and more reliable. Besides, the jobs, turf, positions, and livelihood of these people depended on these ballistic fire bricks. (We were to see similar opposition to the SSTO surface in 1993.) In some ways, Dyna-Soar stretched the existing technology. However, any good development project should do that.

Bono went to work for Douglas Aircraft Company, later to become McDonnell Douglas, where he was technical director for the firm's participation in the NASA Apollo and Voyager space programs.

It was at Douglas that Bono found the creative environment that allowed him to father the SSTO. As the director of Unconventional Launch Vehicle studies and then of Large Launch Systems, he got deeply into all phases of large, reusable rocket design studies. His initial work suggested recovering and reusing the three stages of the huge and expensive Saturn V moon rocket. Then he discovered that the third stage, the Douglas-built S-IVB, was actually light enough to be used as an SSTO capable of putting an 8,000-pound two-man Gemini space capsule into orbit. Bono revised the design so it could be recovered on land with a vertical, rocket-powered landing. He called it "Saturn Application Single-Stage-To-Orbit" or SASSTO.

"Experts" still claim that it's impossible to build a reusable, recoverable SSTO space launch vehicle with a propellant fraction of 0.90 or better. This is a spurious objection because Figure 4-3 shows that many expendable rockets built in the last 40 years have propellant fractions better than 0.90. This table was researched by Major Mitchell Burnside Clapp, a USAF test pilot. In 1994, he pointed out that a 17-gram beer can made by the millions from ordinary aluminum alloy, optimized for packaging instead of performance, and containing 357 grams of adult beverage has a "propellant fraction" of 0.9545.

Bono's SASSTO is the direct ancestor of the McDonnell Douglas "Delta Clipper" SSTO proposal of 1991 and the DC-X sub-scale test rocket that was part of the SSTO plan discussed later.

But SASSTO faced an insurmountable obstacle. It was "born thirty years too soon."

MASS FRACTION OF
CERTAIN EXPENDABLE LAUNCH VEHICLES

(Mass fraction = Propellant Weight/Gross Weight)

Vehicle	Propellant Weight (lb)	Gross Weight (lb)	Mass Fraction
Titan II			
Stage 1	260.0	269.0	0.966
Black Arrow			
Stage 1	28.7	31.1	0.922
Saturn V			
Stage 1	4584.0	4872.0	0.941
Titan III			
Stage 1	294.0	310.0	0.948
Titan IV			
Stage 1	340.0	359.0	0.947
Delta 6925			
Stage 1	211.3	223.8	0.944
Atlas E	248.8	266.7	0.933
Saturn V			
Stage 2	993.0	1071.0	0.927
Zenith			
Stage 1	703.0	778.0	0.903
Titan III			
Stage 2	77.2	83.6	0.923
Saturn Ib			
Stage 2	233.0	255.0	0.913
Titan II			
Stage 2	59.0	65.0	0.908
Saturn Ib			
Stage 1	889.0	980.0	0.907
Ariane 5			
Stage 1	342.0	375.0	0.912
Saturn V			
Stage 3	238.0	263.0	0.905
Energia core	1810.0	1995.0	0.907

Figure 4-3

NASA wasn't looking for cheaper access to space but ways to keep filling the rice bowls of its huge technological army. In 1969, NASA proposed a manned mission to Mars with a space station and a reusable space shuttle as initial elements.

First, the Mars mission was nixed by Congress because getting to Mars wasn't the national security issue like beating the Soviets to the Moon. The Soviets couldn't do the Mars mission at all. Now we know the Soviets tried the manned lunar landing mission and failed in 1969 when their Saturn-sized booster, the N-1, kept blowing up in flight.

Then the space station portion went away. In spite of the successful Skylab space station, the "real" NASA space station slowly withered over the years as dirksens were spent on it. Great forests have perished because of the NASA space station program. It created a lot of business for paper companies because it generated tons of reports. It also kept plywood manufacturers busy because nearly all the hardware NASA has to show for the dirksens spent on the space station is a collection of wooden mockups. Americans once had a space station called Skylab that was allowed to crash after nine astronauts had used it on three missions. A spare Skylab now rests in the National Air and Space Museum and some mockups can be seen in places like Space Center Houston. This caused some astute observer to remark that NASA has certainly been extremely successful in creating expensive museum exhibits.

The only surviving element of the original Mars program is the NASA space shuttle that now costs more than a dirksen every time it's flown.

This environment didn't deter Phil Bono from pursuing the SSTO concept. He knew in his gut that it was the right thing to do in the long run. Nor did it deter others from picking up this concept and trying to do something about it.

FIVE

ROMBUSES and Plug Nozzles

IF YOU'VE BEEN TO DISNEYLAND in California during the period of about 1975 to 1995 and taken the "Flight to Mars" ride, you got a quick glimpse of one of Phil Bono's spaceships. It's an early study called ROMBUS—Reusable Orbital Modular Booster and Utility Shuttle. Although the ride was designed around it, Bono never intended ROMBUS for a trip to Mars. In fact, the Disney people had to invoke a bit of science fiction in the form of a "hyperdrive" to get it there and back. Nevertheless, one part of the presentation shows a ROMBUS sitting on the Saturn V launch pad at the Cape.

Many people thought that ROMBUS was real. It wasn't, of course, although Douglas once made a huge model of it and posed Phil Bono next to it.

ROMBUS wasn't a true SSTO. It featured eight detachable liquid hydrogen tanks positioned around a conical vehicle core. The tanks were expendable and were jettisoned in pairs as their liquid hydrogen fuel was used up.

Furthermore, ROMBUS was a big spaceship although it would have stood only 95 feet high with a base diameter of 80 feet. Liftoff weight was pegged at 14 million pounds. It was designed to lift a million pounds to low-Earth orbit.

The first problem with ROMBUS was that it was too big.

Too big? Hasn't it always been the goal of astronautics engineers to build a rocket that will carry the biggest load? That is Apollo thinking, of course. Size and weight-lifting capabilities may

FIGURE 5-1: *Phil Bono stands beside a scale model of his ROMBUS SSTO at a 1967 travel show in Dallas. (Douglas photo from Stine archives.)*

be technical goals, but they are not necessarily economical business goals. Here's why:

Douglas Aircraft Company didn't build the DC-10 before the smaller DC-3. In fact, Donald W. Douglas, Sr., really didn't want to build the DC-3 for a single requesting customer, American Airlines, because he thought that a 21-passenger airliner was too big for the marketplace. He grumbled, "So they buy twenty of our ships. We'll be lucky if we break even!" Thirty years later, the Douglas DC-10-10 carried more than ten times the number of passengers as the 21-passenger DC-3. But in 1936, an airliner capable of carrying the 222 passengers of the DC-10-10 could never have been filled to the break-even flight load factor.

A million pounds is 500 tons. Even today and into the foreseeable future, the market won't support a spaceship with a payload of 500 tons per flight. Most Earth-to-LEO payloads are in the range of 5,000 to 20,000 pounds—2.5 to 10 tons. In order for ROMBUS to be operated profitably even with a 60% load factor, it would have to wait until 30 10-ton payloads could be gathered for a single launch.

Engineers believed then, and many still do now, that it would be easier to build a large SSTO. This is a questionable assumption given today's technology. The "economy of size" principle can often be misleading. In rail transportation, huge freight cars aren't built; loads are designed to be shipped on standard freight cars and with dimensions that can be accommodated by the existing "loading gauge"—the maximum dimensions of the load that can pass through existing tunnels and over existing bridges. In like fashion, we rarely see very large trucks on the highways. 99.99+% of all loads are carried in the ubiquitous 18-wheeler semitrailers whose loads are, in turn, dictated by the size of underpasses, highway tunnels, bridges, etc.

There is no loading gauge in space. Therefore, spaceships capable of orbiting 500 tons will probably be built someday. In the meantime, the market appears more capable of supporting spaceships with modest payload capabilities in the range of two to ten tons. In the foreseen SSTO operational environment, a lot can be done in this weight range.

The second shortcoming of ROMBUS—and of every Phil Bono spaceship design—was the propulsion system.

Bono quite rightly determined that the best spaceship rocket engine design would be one called a "plug nozzle" or "aerospike."

Only experimental models of such rocket motors have been built and tested since 1960.

The reason why a regular rocket engine isn't the best choice for an SSTO is critical to further discussion of SSTO spaceships.

In nontechnical terms, the Earth's atmosphere has a profound effect on the performance—thrust and efficiency—of a rocket motor.

A rocket engine works best in the vacuum of space.

"But rocket engines have been working well in the atmosphere and in space for decades!"

Not really. Rocket engines have been engineering design compromises.

The rocket engines in the first stage of the Saturn V moon rocket were designed to operate best in the atmosphere. The rocket engines in the upper stages were designed to work best in the vacuum of space.

The Space Shuttle Main Engines (SSME) in the tail of the space shuttle Orbiter are even more compromised because they must work in both environments.

A rocket engine designed to work best in the Earth's soupy atmosphere will not work as well in the vacuum of space. And a rocket engine designed to operate best in outer space won't work well on the surface of the Earth. All of this hinges upon the reason rocket engines have nozzles.

In a rocket engine, propellants are burned in the combustion chamber. This produces large volumes of gas at pressures up to 3,000 pounds per square inch. In order to convert this energy of pressure into energy of thrust or motion, the gas must be expanded by means of a diverging nozzle so that, as it leaves the end of the nozzle, it has the same pressure as the surrounding medium—air on Earth, the near-nothingness of vacuum in space.

The nozzle of any rocket engine converts low-speed high-pressure gas whose molecules move in random directions into high-speed low-pressure gas with all the molecules moving in the same direction.

A rocket engine designed for use at or near the Earth's surface with a surrounding air pressure of 14.7 pounds per square inch has a short nozzle. It needs only reduce the gas exit pressure to that of the surrounding atmospheric pressure.

A rocket engine designed for use in the near-vacuum of space has a longer nozzle. The gas must be further expanded until it exits the nozzle at nearly the pressure of the vacuum of space.

Designing a rocket engine to operate with the highest possible efficiency "all the way" from the Earth's surface to low-Earth orbit poses severe technical problems.

A rocket engine with a large nozzle for use in space will be "over-expanded" when it is operated on the ground and will produce only about 60% of its vacuum thrust.

In a like but reverse manner, a rocket engine designed for use in the atmosphere will be "under-expanded" when operated in space. It doesn't have enough nozzle. Its vacuum thrust will be only about 60% of what it is on the ground.

Given the state of the art in materials, propellants, structures, and fluid dynamics in the 1990s, you "can't get there from here" using a rocket engine designed for best performance on Earth or one designed for space. The efficiency margins are too slim, especially in SSTO spaceships built with present day knowledge.

So an SSTO designer must choose from among several solutions to the problem:

(a) Design a spaceship with multiple rocket engines, some under-expanded and some over-expanded. Shut down the inefficient under-expanded rocket engines as the spaceship leaves the atmosphere, grows lighter because it burns off fuel, and thus requires less thrust.

(b) Use a rocket engine with a nozzle extension that slides backward at the proper time in flight. This nozzle extension converts the engine from being under-expanded to being a space engine.

(c) Use a new type of rocket engine with a "plug nozzle."

Solution (a) means carrying along the weight of rocket engines that aren't used during part of the flight.

Solution (b) means depending upon complex mechanical gadgetry operating in a very noisy and possibly very hot area at the base of the SSTO.

Solution (c) means using a rocket engine concept that hasn't been flown yet.

Bono opted for the plug nozzle concept. The operation of this rocket engine is akin to black magic.

A plug nozzle rocket engine consists of either an annular, toroidal (doughnut-shaped) combustion chamber or a series of small rocket combustion chambers arranged in a ring. In the center of this ring of combustors is a strangely-shaped "nozzle."

Actually, it's only half a nozzle. The other half of the nozzle is created by the presence or absence of the surrounding atmosphere itself.

Figure 5-2 will help explain this.

With a conventional nozzle, operation at the design point of the nozzle (optimum expansion) allows nearly all of the combustion gases to be directed straight to the rear.

If the nozzle is over-expanded, the gas flow separates from the nozzle inside the nozzle and forward of the nozzle exit. The loca-

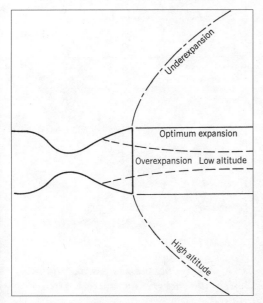

FIGURE 5-2: *Why an ordinary bell-nozzle rocket engine loses efficiency because of under- and over-expansion of the nozzle. (Drawing adapted from Sutton by G. Harry Stine.)*

tion of this separation point can shift back and forth, causing combustion instability and unwanted vibration in the rocket engine. This, in turn, can create problems in the spaceship itself.

If the nozzle is under-expanded, the gas flow blooms outward at the nozzle exit, and all the gas velocity isn't directed toward the rear as it should be for best performance.

With a plug nozzle as in Figure 5-3, the inner plug forms half of an annular nozzle. The surrounding atmosphere forms the other side of the annular nozzle. At low altitudes where the atmospheric pressure is highest, the exhaust jet thus hugs the inner or plug wall, giving the "nozzle" the best expansion under the circumstances. As altitude increases and atmospheric pressure decreases, the exhaust blooms outward, again achieving the optimum expansion ratio.

In essence, a plug nozzle is selfcorrecting. It automatically has the best expansion ratio.

The big problem is that only one or two plug nozzle rocket engines have been designed and tested on the ground. However,

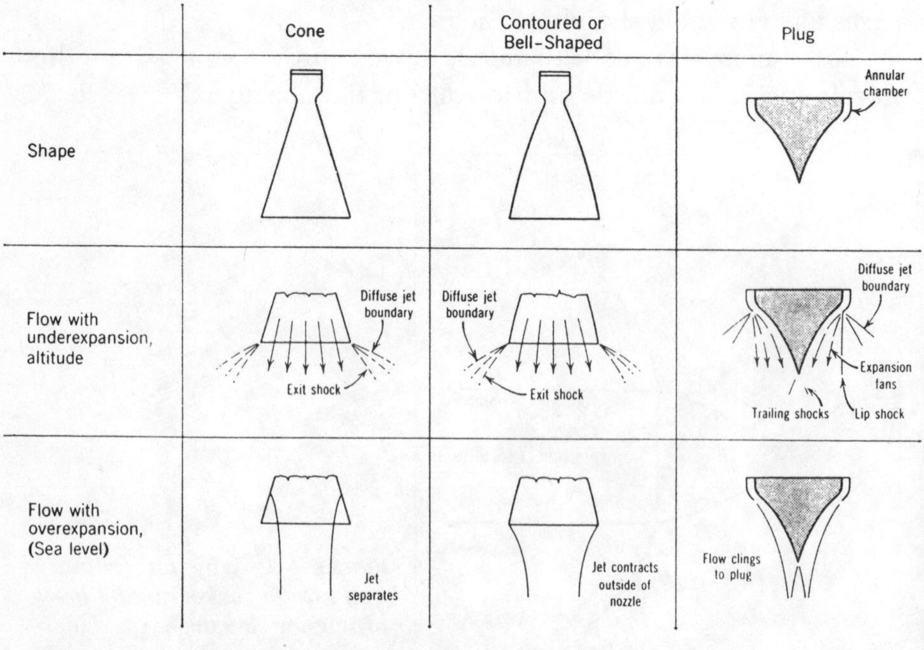

FIGURE 5-3: *How a plug-nozzle rocket engine operates so the surrounding environment itself serves to optimize the nozzle expansion. (Drawing adapted from Sutton by G. Harry Stine.)*

the static tests and the wind tunnel tests showed that the concept is valid.

Phil Bono designed around such a plug nozzle rocket engine although, in the 1960s, only one experimental model had been ground tested by the Rocketdyne Division of North American Aviation, now Rockwell International. In 1971, Rocketdyne built and tested a linear plug nozzle rocket engine that has been proposed for use on the Lockheed-Martin SSTO that will be discussed later.

Plug nozzle rocket engines will undoubtedly be used on SSTO spaceships of the future because the plug nozzle is an elegant solution to the problem of maintaining optimum propulsion efficiency.

However, someone must fund the development of such a rocket engine. NASA and the present-day aerospace contractors operate with the paradigm that the development of any rocket engine requires five years and five dirksens. In the nongovernment commercial milieu, this can probably be done in three years for about $500 million. No rocket engine company can afford to do this in the absence of a military requirement for such an engine. And in this regard, we can rely on the experience of the aviation industry with respect to aircraft engines.

For example, in 1919, appalled by the weight and unreliability of existing airplane engines, Rear Admiral William Moffett, then Navy air chief, issued a requirement for a reliable, light-weight, air-cooled 200-horsepower radial aircraft engine to power the Navy's new fighter aircraft that would fly from the first aircraft carrier, the U.S.S. *Langley*. Curtiss-Wright produced the first of the famous Whirlwind and, later, the improved Cyclone engines.

The demand for radial air-cooled aircraft engines was so large that in 1925 Fred Rentschler left Curtiss-Wright and bought the non-aviation Pratt & Whitney Tool Company, converting it into the company that produced the competing Wasp engine. The Curtiss-Wright Whirlwind air-cooled radial aircraft engine allowed Charles Lindbergh to fly the Atlantic in a single-engine airplane and made possible the classic Douglas DC-2 and DC-3 transports. Cyclones and the Wasps powered nearly all of the military airplanes and the commercial airliners until the advent of the jet engine after World War II.

Again with the jet engine, history repeated itself. Out of the military requirement for a jet engine to propel fighters and bombers, General Electric, Curtiss-Wright, and Pratt & Whitney developed a series of engines that were the ancestors of those powering today's jet airliners. The military requirements that gave birth to the Pratt & Whitney J57 jet engine allowed its civilianized version, the JT3, to be used for the Boeing 707 and other jet airliners.

Will a military requirement for a plug nozzle rocket engine also lead to the civilianized version for the SSTOs? Perhaps. This hinges upon the realization by the Department of Defense that commercial space activities and facilities will require the same sort of protection as their earthbound counterparts and oceangoing trade. And when the civilians who run the defense of this nation realize that space power is now just as important for the national defense as air power and sea power.

Phil Bono went on to design an intercontinental passenger and troop carrying rocket, the Pegasus and Icarus respectively. Both used plug nozzle rocket engines.

Although Bono figured out that the Saturn S-IVB stage could be made into an operable SSTO with an 8,000-pound payload, he was operating under the then-current paradigm: Size equals economy. The bigger, the better. A million pounds to orbit was the goal. As a result, given the technology of the time, Bono couldn't come up with an SSTO that would do the job without throwaway tanks.

Or without a rocket sled launch to take the place of a lower stage, thus increasing the payload by 20%. Bono proposed using a rocket sled climbing up the face of a 5,000-foot mountain and releasing the Hyperion SSTO at maximum sled velocity.

Launching a spaceship from a rocket sled wasn't a new idea. During World War II, two Austrian rocket pioneers, Dr. Eugen Sänger and Dr. Irene Sänger-Bredt, conceived an intercontinental rocket plane launched by a rocket sled.

But mountainside sled launching requires that spaceports be located where mile-high mountains exist. Given present environmental consciousness, a mountainside rocket sled might raise the ire of environmentalists intent upon preserving pristine mountain wilderness areas. In addition, the concept introduces difficult engineering problems into the system, violating the basic principle of good, efficient, economical engineering: KISS (Keep It Simple, Stupid!).

I proposed writing a rocket transport book with Phil Bono in 1967. We got as far as a book outline and a few sample chapters. Then I ran into a show stopper: the economics of rocket transportation. Realizing that rocket transportation had to follow the same basic business principles as airline, ocean, truck, and rail transportation, the two of us admitted that it was difficult to make a case for rocket transportation *at that time.* I'd forgotten Cornog's paper and tried with Bono's help to get some economic, production, and operational data on airliners and airlines. What Bono found in Douglas archives and among the company's economists wasn't sanguine for the rocket transport case. Douglas had some very good numbers on aircraft production and airline operating costs. No matter what I did with the numbers, a 1967 sled-launched SSTO did not compute as economical.

Today we do indeed have a case for the simple SSTO. Therefore, this is probably the modern version of the book Phil Bono and I wanted to write more than a quarter of a century ago.

Bono later teamed with Kenneth Gatland to write *Frontiers of Space*, published by Blandford Press in London in 1969. Although it was a "space is a nifty thing" book typical of the period, it did get Bono's concepts into print.

However, none of the above detracts from the fact that Bono was there first, leaving his stake in the ground to mark the start of the trail for others to follow. Despite the fact that the post-Apollo cutback caused Douglas to reassign Bono to designing hydraulic actuators for the DC-10 and MD-80 programs (which he did until he retired), others did follow.

SERVS and
Space Merchants

BETWEEN 1969 AND 1971, many proposals for reusable SSTOs were generated during the space shuttle feasibility study. NASA managers and engineers were trying to determine how to carry out the next big goal they'd created internally in order to justify the continued employment of the technical army organized in 1961 for the manned lunar landing program.

A lunar base seemed to be the next logical step. However, logical steps didn't seem to be in fashion following the Soviet launch of Sputnik 1. If we had entered space logically, we would have done it differently.

How? If in 1956 anyone had asked any of the space visionaries how we ought to go to the Moon, the answer was almost universal:

First, build a true earth-to-orbit space transportation system with reusable spaceships.

Then build one or more space stations in orbit around the Earth. These would be used to assemble and fuel spaceships expressly designed to operate only in the vacuum of space and used to transport people and cargo to the Moon.

What the United States government did in the Apollo program was quite different. Apollo was basically a program to send men directly to the Moon and bring them back to Earth using the ballistic missile technology then available. This was the task given to NASA by President John F. Kennedy in May 1961.

BALLISTIC VERTICAL TAKEOFF AND LANDING SSTO'S
HISTORICAL PERSPECTIVE

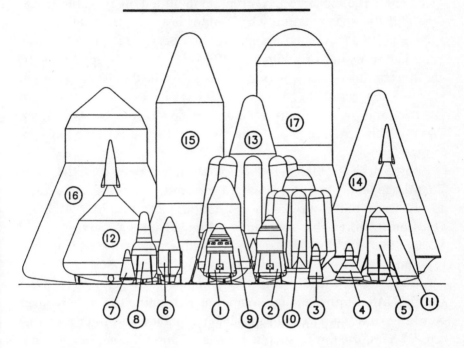

1) Phoenix E (1985) PacAm
2) Phoenix C (1982) PacAm
3) Phoenix L (1974) Hudson
4 Phoenix L' (1976) Hudson
5 S-IVB (1965) Douglas
6) SASSTO (1967) Douglas
7 ATV (1972) NASA Marshall
8) BETA (1969) MBB-Germany
9) Hyperion (1969) Douglas
10) Pegasus (1969) Douglas
11) Unnamed (1970) NASA-OART-MAD
12) SERV (1971) Chrysler
13 ROMBUS (1964) Douglas
14) NEXUS (?) Ehricke
15) Unnamed (1983) Rockwell
16) Unnamed (1977) NASA-JSC
17) "Big Onion" (1976) Boeing
18) RITA (1961) Douglas (not shown)

Figure 6-1: *A comparison of historic SSTO proposals. (Drawing courtesy Tom A. Brosz and Pacific American Launch Systems, Inc.)*

There was never any intent on the part of Kennedy or his space advisor and chairman of the National Space Council, then-Vice President Lyndon B. Johnson, to establish a lunar base, much less a space station.

NASA's 1969 future space program included the reusable spaceship and the space station. On Wednesday, July 16, 1969 shortly after Apollo-11 was launched to the Moon, Arthur C. Clarke remarked to Walter Cronkite on CBS News TV from Cape Kennedy, "In this coming ten-year period we've got to see the development of reusable spacecraft. . . . We've got to have spaceships that we can use over and over again as often as we use a conventional airliner. The reusable space transporter is the next thing which we have got to get."

But Clarke implied what everyone else accepted at the time: That the federal government would continue to explore and exploit space. Even the Kubrick-Clarke motion picture, *2001: A Space Odyssey*, mirrored this government-directed space program.

By 1969, the commercial space paradigm of the 1950s had been lost and forgotten. We've had to reinvent it in the 1990s.

Fresh from winning the technological war with the Soviet Union, NASA upper management was determined to do whatever would ensure the survival of their bureaucracy and their standing army. Therefore, they issued contracts to study the various technological options available to them for a follow-on: a reusable space shuttle. The mission requirements, including the dimensions and weight of the required shuttle payload, were based on building a space station. At that time, as is still the case today, NASA managers thought in terms of "missions." Airline executives, on the other hand, think in terms of "flights."

One of the study contractors was Chrysler Corporation's Space Division in New Orleans, Louisiana, that built the massive first stage of the Saturn V moon rocket. These Chrysler engineers certainly had ample expertise in the design and construction of large space vehicles.

Working under contract NAS8-26341, Chrysler produced a six-volume report on June 30, 1971, covering their design and analysis of an SSTO called the Single-stage Earth-orbital Reusable Vehicle, or SERV. Because people did the work, not computers, people should get the credit. Those involved were: Charles E. Tharatt, Study Manager; William R. Baldwin, Principal Systems Analyst; John H. Wood, Principal Performance Analyst; and Arthur P. Raymond, III, Principal Program Analyst. The NASA study manager was Robert J. Davies. Readers who might like to read the documentation can reference the contract number as well as the designations on the report: DRD MA-095-U4 and TT-AP-71-4.

SERV looked like a Phil Bono design.

SERV had an Apollo capsule-like shape, was 100 feet high, and had a base diameter of 96 feet. Its cargo hold was 15 feet in diameter and 60 feet long (the same size as the later space shuttle Orbiter's payload bay). It was designed to take 88,000 pounds to orbit and return. Featuring vertical takeoff and vertical landing, SERV was totally reusable. However, instead of using rocket propulsion for landing, it carried a separate system consisting of jet engines and ordinary jet fuel. Its rocket propulsion system was a huge plug nozzle unit with 12 propulsion modules arranged around it. Propellants were liquid hydrogen and liquid oxygen. The rocket propulsion system would produce 7,450,000 pounds of thrust at liftoff. SERV was designed for a vehicle life of 10 years and 100 flights.

Although NASA didn't buy the SERV design, it is of more than passing interest to note an important fact: *In 1971 with the technology, materials, and propulsion then available, Chrysler Corporation is on record as the designer of an SSTO with a propellant fraction of 0.918.*

More than 20 years later, some NASA managers and engineers were still vehemently denying that it was possible to achieve the 0.90 payload fraction required for an SSTO to be feasible!

SERV wasn't the only vertical takeoff, vertical landing (VTOVL) SSTO studied, by the way.

Space visionary Krafft Ehricke came up with one called NEXUS. Out of Messerschmitt-Bolkow-Blohm in Germany in 1968 came BETA. Even NASA itself in 1972 studied one called ATV while yet another unnamed design came from NASA Johnson Space Center in 1977.

But the person who really carried the torch for the SSTO during the 1970s and the early 1980s was a young man from St. Paul, Minnesota, who had read Phil Bono's papers and reports. He wanted to build spaceships.

Gary C. Hudson still wants to build spaceships. He decided along the way that going into space permanently could be done only if someone—including himself—could make a buck doing it.

I may have had a role in that. At a meeting of the American Astronautical Society at the San Francisco Airport Hilton Hotel in October 1977, I presented two papers on space transportation marketing and economics. I told the assembled starry-eyed space advocates and tech-loving rocket engineers, "Stop talking to yourselves and learn how to speak the language of people who have the money to pay for what you want to do—and I don't mean the federal government." Gary Hudson was one of those who listened to me and tried to do just that. He ceased being a pure space advocate and became one of the first space entrepreneurs.

However, many old-time space advocates and pioneers thought Gary Hudson was a bothersome dilettante. "Real space engineers" looked upon his desire to make money in space as somehow distasteful. Hudson was polluting the pure dream. As such, he was labelled by some people as a con artist or worse because he wanted to use investors' money rather than tax dollars. Even today, many space advocates and rocket engineers continue to pursue the tax dollar as a source of funding. If they'd put as much effort into raising private funds as they have into the circus politics of the federal budget process each year, we might have had private spaceships ten years ago. Or maybe not, as we'll see.

Some space advocates believed Hudson had stolen Phil Bono's designs and passed them off as his own. But Hudson was and is not a dilettante. Although not possessing an engineering degree, he picked up the true SSTO concept from Bono and proceeded from there.

Hudson formed Foundation, Inc. and its subsidiary, the Foundation Institute, a nonprofit space think tank. The young commercial space advocates involved with Gary Hudson in Foundation, Inc., were Tom A. Brosz, Rick Sternbach, and Charles Duelfer.

Hudson discovered he couldn't raise any money to make SSTO happen in the 1970s. However, he certainly did raise and maintain the interest level in SSTO during the long decade when every other space dreamer was counting on the cheap, regular, reliable, and scheduled service deceitfully promised by NASA for the space shuttle.

When it became obvious to him that the non-profit Foundation Institute couldn't attract capital, Hudson formed the for-profit corporation, Space Merchants, Inc., on May 30, 1974, with Gary R. Lindberg and Stanley E. Williams.

To this day, Gary Hudson has remained firm in his commitment to making profitable commercial space transportation possible. He qualifies as a true revolutionary because he committed his life, his fortune, and his sacred honor to the goal.

What Hudson and his colleagues did is important in the history of the SSTO. They designed an SSTO named Osiris, a Vertical-Take-Off-Vertical-Landing (VTOVL) spaceship capable of carrying 24,000 pounds of payload to orbit. It would have weighed 56,000 pounds empty and 640,000 pounds ready to fly.

What is important about Osiris is that the payload compartment was situated between the liquid hydrogen tank in the nose and the liquid oxygen tank below it. This was a radical departure from the basic artillery shell design of rockets where the payload was put in the nose, just like an artillery shell. And it was the first time this configuration was used. Not even Phil Bono departed this far from classical rocketry. Keep Osiris in mind because it's the direct ancestor of the McDonnell Douglas Delta Clipper design of 1991.

The Osiris design was followed by Phoenix, which was smaller— 58 feet high and 31 feet across the base with a gross takeoff weight of 328,000 pounds. It was designed to carry 10,000 pounds to orbit.

Again, keep these numbers in mind. They are about the same as those of the Delta Clipper and other SSTO spaceship designs of the 1990s. Hudson was the first to realize that it wasn't necessary to build a huge space-going version of an oil tanker when an 18-wheeler semitrailer truck would do the job just fine.

Hudson saw another space launch market: small satellites. So Space Merchants scaled-down the Phoenix to create the ATV. This would have been 34 feet tall with a 16-foot base diameter, a gross weight of 53,000 pounds, and the ability to place 1,100 pounds into orbit.

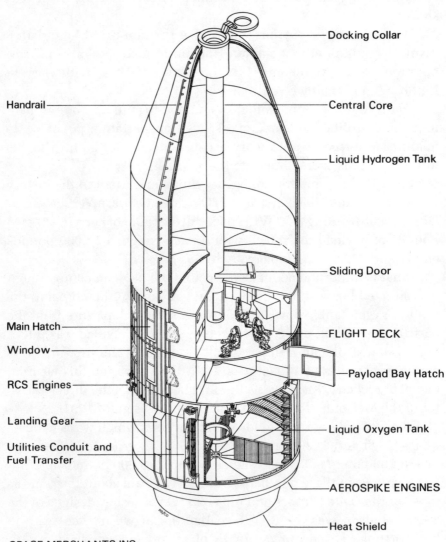

Docking Collar

Handrail

Central Core

Liquid Hydrogen Tank

Sliding Door

Main Hatch

Window

FLIGHT DECK

Payload Bay Hatch

RCS Engines

Landing Gear

Liquid Oxygen Tank

Utilities Conduit and
Fuel Transfer

AEROSPIKE ENGINES

Heat Shield

SPACE MERCHANTS INC.

FIGURE 6-2: *Cutaway drawing of Gary Hudson's OSIRIS SSTO, almost a direct ancestor of the McDonnell Douglas Delta Clipper concept. (Drawing from Space Merchants, Inc., and G. Harry Stine archives.)*

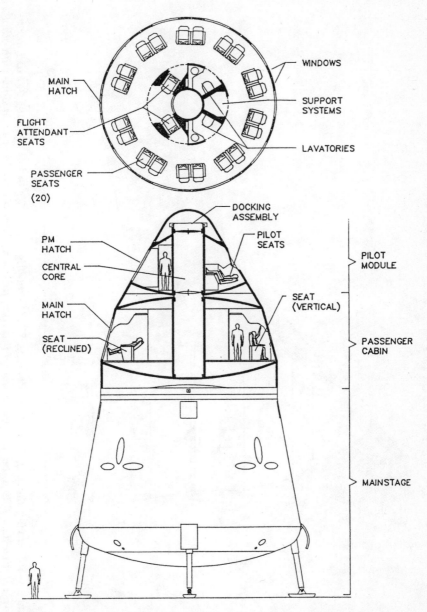

FIGURE 6-3: *Cutaway drawing of the PacAm Phoenix SSTO that Society Expeditions wanted to use for space tourism. (From Pacific American Launch Systems, Inc., and G. Harry Stine archives.)*

COMPARISON: 747 VS. PHOENIX

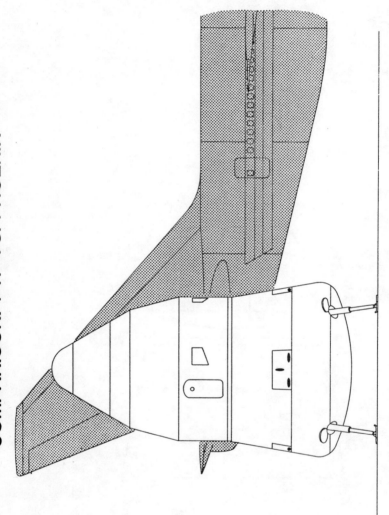

FIGURE 6-4: *Drawing showing the PacAm Phoenix SSTO standing next to the tail of a Boeing 747 airliner to show relative size. This is the first apparition of this comparison that has been widely used since with other SSTO proposals. (From Pacific American Launch Systems, Inc., and G. Harry Stine archives.)*

Osiris, Phoenix, and ATV form a line of direct ancestry for today's SSTO spaceship designs. Remember, this work was all done privately by Hudson and his associates in the 1970s. They were trying to break the paradigm.

But Gary Hudson had never built a rocket. He was snubbed by the experienced rocket engineers who had. In the 1970s and early 1980s, none of them believed Gary Hudson could do it or that his Space Merchant designs were feasible.

He was soon given the chance to build a rocket, but not the SSTO he wanted.

SEVEN

Space Services and Society Expeditions

THE REUSABLE SSTO appears to be the viable and attainable solution to lower-cost space transportation today. However, some engineers believe another solution is called the "Big Dumb Booster," a large, relatively cheap, extremely crude, and very "stupid" (from the viewpoint of its autopilot and other control devices) expendable single-stage-to-orbit rocket. The Big Dumb Booster is an attempt to get rid of all the extraneous bells and whistles required by the military services for ballistic missiles and produce a simpler space launch vehicle.

Without going into the details of how it started, in 1980 Gary C. Hudson—now relocated in the San Francisco Bay area under the company name GCH, Inc.—was given a contract by Space Services, Inc. of America to design, build, and test a cheap, simple, expendable satellite launch vehicle. It was Gary Hudson's version of a Big Dumb Booster.

It was perhaps the first privately-funded space launcher project in the United States.

Behind Space Services, Inc. of America was David Hannah, Jr., a Houston real estate magnate and land developer.

In the first phase of the project, GCH, Inc. was to build and test the basic core module of a rocket Hudson called Percheron because it would become the work horse of space transportation. The basic module using liquid oxygen and kerosene-like RP-1 fuel was to serve as the foundation for more powerful rockets made up of clusters of this module. Each module was propelled by a single uncooled rocket combustion chamber producing a thrust of 60,000 pounds for 25 seconds. The rocket engine used a simple "pintle injector" that was 100% successful in the ascent engine of the Apollo Lunar Module (LEM). No turbopump was used; the propellant tanks were pressurized with helium to force the rocket propellants into the combustion chamber.

The Percheron was about as simple as a liquid propellant rocket can be. The first basic flight test module would not incorporate any guidance system; it would be a fin-stabilized rocket like the millions of model rockets flown by hobbyists. But it was no small rocket. The basic module was 44 feet long and 4 feet in diameter. When launched from sea level, it could attain a maximum altitude of 373,000 feet 164 seconds after takeoff.

Hudson sent me some drawings of Percheron in early 1981. I looked at them and gave him a call at once. I told him the Percheron wouldn't be stable in flight.

Hudson asked me why, and I told him: not enough fin area. In fact, no fins at all, only open tubes arranged around the base of the rocket. These don't work at supersonic speeds. Percheron would reach the speed of sound at an altitude of about 4,000 feet some seven seconds after lift-off. It's maximum velocity would be about Mach 5.

Hudson asked me to help, which I did under a consulting contract in March 1981. I told him to put four sheet aluminum fins on the Percheron and also told him how big they ought to be. The numbers came from a computer program developed to ensure that model rockets would fly straight when built and flown by hobbyists. It worked for models and it would work equally well for Percheron.

Far more fascinating than the technology behind Percheron, however, was the reaction from the professional rocket engineers, especially those working with and for NASA. Publicly, they pooh-poohed Percheron. So did the national news media. *Time* magazine in its June 29, 1981 issue, wrote, "In Ian Fleming's novel, *Moonraker*, Multimillionare Hugo Drax built himself a huge rocket to annihilate the city of London. He was foiled in his sinister stratagems by James Bond, Agent 007. Now a businessman space buff named Gary Hudson is trying some rather far-out capitalism of his own, with a plan to start putting satellites into orbit from a private launching pad in Texas by 1983. So far, not even the U.S. Government is trying to stop him."

But the United States government did try. Rockets are considered to be munitions. So the feds tried to stop the launch by claiming that GCH, Inc. and Space Services, Inc. were exporting munitions because the Percheron would be launched from Matagorda Island into the international waters of the Gulf of Mexico. Thus, it would be "exported" out of the United States. If this sounds bizarre, you haven't had to deal with the federal bureaucracy. Several government agencies with different regulations got into the act. It was only the work of Houston lawyer and space buff, Arthur M. Dula, that saved Hudson. Dula had to obtain numerous permits and licenses merely to fly a modest suborbital rocket.

A contemporary editorial cartoon in the *Chicago Tribune* took a poke at the Percheron. It showed three fat, bald men in business suits observing the Percheron launch site. One of the men is saying to the other two, "Well, it's settled, then. When it comes to our first moon shot, our crew will be Ronald McDonald, Mister Clean, and the Jolly Green Giant!"

NASA and aerospace contractor managers watched in silence. Some didn't think Percheron would work. Others feared that it would. They were ready for whatever happened.

At 4:57 P.M. on August 5, 1981, a static test of the Percheron rocket motor was begun. The Percheron was bolted to its launch pad. The test was to be only a five-second burn to check out everything.

However, the humid Gulf air had caused ice to form on the cold liquid oxygen valve, literally freezing it closed. The valve didn't thaw out until the rocket engine was ignited. The failure of the rocket propellants to get to the combustion chamber at the right time in the right amounts produced what rocket engineers call a "hard start" or a "catastrophic disassembly."

In short, the Percheron blew up on the launch pad.

The upper part of the rocket went about 250 feet into the air, not high enough to beat the altitude record for a static test of five miles set by the U.S. Navy Viking rocket #8 in 1952. Because Matagorda Island was prime grazing land for a very large herd of cattle, the

Percheron explosion started the largest cow chip fire in Texas history. Fortunately, the alligator that lived in the pond used to collect the launch stand cooling water survived.

Every rocket engineer has received the baptism of fire caused by a hard start. It's part of the rocket business. "Rocket scientists" never have hard starts because they never build and test rockets; they only design and write reports about them. Rocket engineers do all of the above.

But people at NASA's nearby Lyndon B. Johnson Space Flight Center in Houston used the Percheron explosion as their excuse to do what they believed they had to do to save the space shuttle and the NASA bureaucracy.

They went at once to David Hannah, Jr., whose investment had just blown up. They offered to "help."

Their "help" resulted in Hannah showing up at the GCH factory a few weeks later with a check and a contract cancellation. Hudson was out of the Big Dumb Booster business with Space Services, Inc.

Several months later, several managers and engineers resigned from NASA and joined Space Services, Inc. A few months after that, the new Space Services group launched the solid-propellant Conestoga using a surplus Minuteman ballistic missile rocket motor "borrowed" from NASA under a deposit refundable when the borrowed motor was returned. Space Services forfeited the deposit, of course, because the borrowed rocket motor ended up exported to the bottom of the Gulf of Mexico.

Strange as it may seem, this actually happened because Space Services couldn't buy the surplus Minuteman ICBM motor! It was government property released by the U.S. Air Force for NASA use.

It's possible to do strange and wonderful things when you're connected to a government bureaucracy where they write their own rules. All they had to do was keep the paperwork clean.

Conestoga was the last rocket Space Services could afford because the ex-NASA engineers had approached the project with the usual government paradigm: Cost is not a factor.

Furthermore, after Conestoga, the ex-NASA people at Space Services really didn't want success.

I can and will report here exactly what went on at NASA Johnson Space Flight Center because I was told by someone who was involved. It has a definite bearing on what happened to the SSTO program in 1994.

My source is Dr. Maxime A. Faget, one of the people who led NASA's manned space flight program—Mercury, Gemini, and Apollo. Faget also helped determine the final configuration of the NASA space shuttle. By 1981, the space shuttle had flown for the first time, years behind schedule, and NASA was eagerly looking for commercial shuttle payloads. Space Services, Inc. was a competitor. NASA managers knew only one way to confront competition: destroy it. And they did.

On Saturday, October 3, 1981, Faget was present at the annual induction ceremonies of the International Space Hall of Fame in Alamogordo, New Mexico. At the reception about 8:00 P.M. that evening in a function room at the Holiday Inn, my wife and I were talking with Faget. I happened to mention my disappointment in the cancellation of the GCH Percheron contract with Space Services.

Faget looked me directly in the eye and stated, "The rocket business is no place for amateurs like Gary Hudson. He had to be stopped and we stopped him."

To Faget, an "amateur" was anyone who didn't work for NASA or who didn't have a government rocket contract. He said so to my wife and me.

Faget went to work for Houston-based Eagle Engineering, Inc., and former astronaut Donald K. "Deke" Slayton joined Space Services, Inc. following their retirement from NASA in late 1981 and early 1982. With former NASA techno-bureaucrats running Space Services, Inc. instead of entrepreneurs, the company slowly ceased to do business. This story is important because it appeared to be repeating itself with another entrepreneurial space transportation company in 1995.

But this didn't stop Hudson who picked up the pieces and went back to work.

GCH, Inc. revived the Phoenix SSTO project and carried on with "negative income."

In September 1984, GCH, Inc., was approached by Society Expeditions, Inc. to provide a spaceship for Project Space Voyage. Society Expeditions, Inc., a Seattle, Washington, operation, was the brainchild of T. C. Swartz. This company would plan, organize, and take care of all the details if you wanted to hold your twenty-fifth wedding anniversary at the North Pole with all of the original wedding party. They'd arrange for everything if you wanted to follow the trek of Stanley searching for Livingstone through deepest, darkest, dankest Africa. Society Expeditions would put together (for a fee) an adventure without adventure, the ultimate in tourism.

Swartz had approached NASA in the early 1980s with the idea of renting a space shuttle Orbiter and installing in its payload bay a passenger module capable of carrying 74 people. He wanted to offer one-day tours to orbit and back for $50,000 a seat.

NASA administrators and managers, of course, didn't believe that this sort of crass commercialism should be allowed to tarnish such a scientific national activity as the space shuttle, much less the space program. The space shuttle, paid for by taxpayer dollars— far more of them than had been anticipated—was intended for use only on missions involving national security and for pristine scientific purposes and experiments. Carrying tourists to and from orbit simply couldn't be tolerated!

In retrospect, NASA's refusal to deal with Society Expeditions may have saved 74 lives.

Swartz decided he'd have to buy and operate his own spaceships. Hudson—whose company had become Pacific American Launch Systems, Inc. or "PacAm"—responded to a query in this regard from Swartz. The initial dealings between PacAm and Society Expeditions began with a proposal from Swartz for PacAm to build and sell Phoenix SSTOs to him.

Hudson, realizing that Society Expeditions probably couldn't keep several Phoenix SSTOs flying often enough to be economical, proposed to lease Phoenix SSTOs to Society Expeditions instead and operate them for other payload flights.

That was the deal that was finally cut in September 1984.

Then PacAm and Society Expeditions went looking for venture capital to do it.

They both ran into a banker's version of a brick wall.

Any investor who remains an investor quickly learns to exercise "due diligence." This includes checking out the business plan and going over the statement of source and use of funds. For a venture that depends on technology not yet widely used, it also means checking out that technology. Most investors aren't interested in the technology and aren't technologists themselves. (How many rich aerospace engineers do you know?) Therefore, they find someone they believe to be familiar with the technology and listen to what that person says. Usually, it's a friend or relative.

Hudson quickly discovered that practically every investor or venture capitalist had a relative who worked for NASA or in the aerospace industry.

When potential investors took Hudson's Phoenix SSTO proposal to their "experts" who worked in or for NASA—and these were the only rocket experts available—the evaluation they received was uniform: "It won't work. If it would work, NASA would be doing it."

This may not be totally fair, because engineers are, by education and training, extremely conservative when it comes to applying technology. The bridge must not fall down. The wing must not come off the airplane. Engineers are not paid to speculate but to design things that must work. In some cases they can lose their jobs and/or their professional licenses if they don't. However, as Henry Petroski, Professor of Civil Engineering at Duke University, points out in his outstanding and readable books on the role of failure in engineering, design, and invention, engineering itself progresses as a result of failure in design while invention is motivated by the failure of existing designs to do what is wanted and expected of them.

In some cases, potential investors will ask a friend at the local university to render an opinion. That's an even worse mistake. By and large—and this comes from my own experience—university professors (and textbooks) are about five years behind any given technology in the world of commerce and industry. Guess what sort of an opinion an academician will render about any proposal involving high technology that's just beyond his (or her) scope of knowledge and expertise, especially when the academic's reputation for being a science guru is on the line to a wealthy investor who might instead donate that money for a new science building or laboratory?

Hudson discovered all this and more.

The lack of money to build the PacAm Phoenix didn't stop Swartz from vigorously promoting Project Space Voyage. Society Expeditions solicited and took reservations for a seat to orbit and back for $50,000 a ticket with $5,000 down. This money was held in escrow to reserve the seat.

Swartz sold 500 reservations within a year.

But he had to give the escrowed money back.

Project Space Voyage became a nonviable commercial service at 11:39 A.M. EST on January 28, 1986, when the space shuttle carrying the Challenger on Mission 51-L blew up, killing all seven people aboard. Some of us wondered what would have been the public reaction if it had been carrying a passenger module with 74 paying passengers as well.

The wonderful spectacle put on by the government with tax money was suddenly seen as deadly. Space was no longer a "beat the Soviets" game. People were reminded that no frontier has ever been opened without the loss of life.

The Challenger disaster marked the beginning of the end of the government space monopoly and the dim dawning of the true commercial space age.

EIGHT

Hunter and
the Council

MAXWELL W. HUNTER IS a real rocket engineer and has been for more than 50 years. He has a long list of outstanding accomplishments as an engineer, project manager, and policy advisor.

Hunter got started with most American rocket pioneers after World War II. Working for Douglas Aircraft Company from 1944 to 1961, he designed the famous Nike antiaircraft missile, the air-to-air Sparrow missile, and the Honest John artillery rocket.

On December 27, 1955, Douglas was chosen to build the Thor Intermediate Range Ballistic Missile (IRBM) to supplement the 5,000-mile Atlas and Titan InterContinental Ballistic Missiles (ICBMs). Thor was to be a smaller, simpler missile using as many Atlas parts and systems as possible and carrying a nuclear warhead. The United States needed Thor to allay the fears of the western European allies who had come under the threat of missile attack by Soviet ballistic missiles. Thors would be based in Great Britain and Turkey.

Douglas put Max Hunter in charge of the Thor project.

Hunter had the first Thor missile on the Cape Canaveral launch pad in January 1957, a bit more than a year from the day the contract was signed. He did it within budget, too. Hunter organized and led a lean, highly-focused crew of engineers and technicians in what would now be termed a Skunk Works operation.

In common with every rocket engineer, Hunter had a spectacular failure when the first Thor missile blew up on the launch pad

FIGURE 8-1: *Maxwell W. Hunter at the fifth flight test of the DC-X, White Sands, June 27, 1994. (Photo by G. Harry Stine.)*

on January 25, 1957. It took four flight tests before Hunter had what he considered a success: on November 26, 1957, Thor delivered its dummy warhead on target 1,500 miles away.

Hunter's Thor grew up and, with upper stages, evolved into the McDonnell Douglas Delta rocket, one of the most used and most reliable expendable space launch vehicles in history.

Hunter then spent three years on the professional staff of the National Aeronautics and Space Council, an advisory group to the President of the United States. Afterward, he went back to the aerospace industry, spending 22 years with Lockheed Missiles and Space Company.

I've known Hunter since his White House days. He's not only an outstanding engineer but a visionary who is quick with a smiling retort, disarming in his directness, cognizant in areas other than mere engineering, and utterly brilliant. We've traded ripostes and shared ideas for years. Hunter is a hair shirt. He asks embarrassing questions and is always about three moves ahead of any debate or conversation. Working with him is both a Class Three Headache and a joy. Usually the latter. I say this with great respect for the man, because he managed to survive and prosper in the aerospace government-industrial environment for more than 50 years without being fired even once, a monumental achievement in itself.

Hunter played a major role in the birth of the SSTO.

He first became interested in the concept at Douglas in 1959 when he was involved in the design of an SSTO called Rita that was propelled by a nuclear rocket engine. With the materials, design techniques, and structural technologies of that time, only a nuclear rocket engine had the efficiency—"specific impulse" in rocket engineering terms—that would permit an SSTO. However, the Nuclear Test Ban Treaty brought a halt to all nuclear rocket engine development in 1963.

In 1973 after evaluating an SSTO proposal by Robert Salkeld and Rudi Beikel, Hunter replied when asked about how close we were to SSTO technology, "Not very."

But Hunter remained interested in the SSTO concept because it made sense to him. He published several technical papers on this

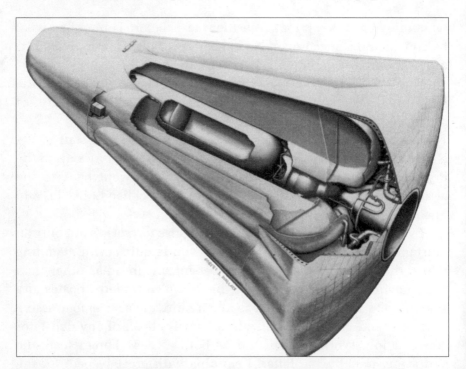

FIGURE 8-2: *Rita, the nuclear-propelled SSTO designed by Max Hunter. (Drawing courtesy Max Hunter.)*

and acted as a consultant to Hudson on the design of the Phoenix that Society Expeditions wanted. When Project Space Voyage came to a halt as a result of the Challenger accident, Hunter—who was still with Lockheed Missiles and Space Company—went to work in 1986 on a new SSTO concept he called the X-rocket.

Hunter designed this X-rocket to use as much off-the-shelf technology as possible. It was to be powered by a cluster of 24 Pratt & Whitney RL-10-35K rocket engines that used liquid oxygen and liquid hydrogen. The RL-10 is probably the most efficient and most reliable liquid rocket engine ever designed and built. The RL-10 was perfected in 1960 and propelled the upper stage of the Atlas-Centaur expendable space vehicle as well as the upper stages of the early Saturn I rockets. It could be started and stopped in orbit. It's still being used today.

The major element of Hunter's proposal was his insistence on eliminating the huge "standing army" of ground personnel needed to prepare and launch a space vehicle. His Douglas Thor missile was launched by a crew of about a dozen military officers and enlisted men. The enormous cost of launching the space shuttle is mostly due to the fact that more than 10,000 people are involved.

The Lockheed X-rocket wasn't an official company project. It was Hunter's brainchild. Because of his years in the business and his acquaintance with many people as a result of his service on the National Space Council, Hunter briefed people as high up as the Secretary of the Air Force. Most listeners were enthusiastic about the concept. However, sooner or later someone asked the Lockheed executives how strongly the company was behind this proposed spaceship. "What spaceship?" was the astonished reply. The Lockheed officers knew nothing about it. So an internal Lockheed evaluation committee was formed. Most of the members were engineers who had worked on the Navy's Trident submarine-launched ballistic missile, a piece of ammunition. Lockheed people apparently believed that a rocket was a rocket, period. After all, both the Trident and Hunter's SSTO proposal were rockets, weren't they?

Hunter didn't win that one. He'd developed the X-rocket proposal on a shoestring. The Lockheed evaluation team had funds and authority that Hunter didn't. The team could call up cadres of engineers who could scrutinize the smallest element of the proposed design. Because a military or naval ballistic missile must be built to specifications that are quite different from those of a spaceship, the results of the internal Lockheed evaluation were predictable: "It can't be done!"

On the other hand, an independent evaluation performed by the Aerospace Corporation said it could be built but that it wouldn't be able to deliver a payload to orbit. It would work, but it wasn't worth the trouble.

The Bell X-1, the first supersonic airplane, didn't carry a payload, either, only an Air Force pilot named Captain Chuck Yeager and a broom handle.

However, the Aerospace Corporation report was the first evaluation of an SSTO since the Chrysler SERV proposal nearly 20 years before. Technology does not run in reverse. SERV could have been built in 1970 with a large payload capability. In 1986, no one but Hunter and Hudson believed it was possible to build an SSTO that would deliver a payload to orbit.

The engineers were obviously welded firmly to their existing ammunition technology.

This was the situation when the Council became involved.

The Council is not the National Space Council. It's the Citizens' Advisory Council on National Space Policy. It was formed late in the evening of Wednesday, November 12, 1980, in an empty conference room in the von Kármán Auditorium at the Jet Propulsion Laboratory, Pasadena, California. Present were Dr. Jerry E. Pournelle, Betty Jo ("Bjo") McCarthy Trimble, my wife Barbara, and me. We had come to JPL for the Voyager I encounter with the planet Saturn.

Jerry Pournelle had worked for Boeing on space suits and other space projects. He'd become a fulltime author and a well-known computer journalist. Jerry was (and still is) into conservative Republican politics. He'd been a speech writer for Ronald W. Reagan as California's governor. He was well connected in conservative circles.

Bjo Trimble is the indefatigable lady who spearheaded the letter-writing campaign to keep Gene Roddenberry's Star Trek television series on the air in 1968. She'd also led the letter campaign to name the first space shuttle Orbiter the *Enterprise.*

FIGURE 8-3: *The Citizen's Advisory Council on National Space Policy meeting on May 10, 1986, in Tarzana, California. Dr. Jerry Pournelle (standing against bookcase) leads the discussion on what should be done following the loss of the space shuttle* Challenger. *(Photo by G. Harry Stine.)*

As for my c.v., check the biography at the end of this book.

Pournelle pointed out that we had a unique opportunity to influence the future space program of the United States because the new Reagan administration was looking for input. Jerry knew the people at 1600 Pennsylvania Avenue. He knew The Man himself. If we could organize a private group of some of the best and most forward-thinking people interested in space, any white papers we produced would be read in the White House.

The ground rules for the organization of what came to be known as the Citizens' Advisory Council on National Space Policy were simple.

Beyond the core group selected by Jerry, Bjo, and me, every potential new member would be invited to join only if no negative comments or objections were received from other Council members.

Council membership grew to include visionary aerospace engineers, lawyers, financiers, and other professionals, plus knowledgeable "hard science fiction" authors such as Robert A. Heinlein, Larry Niven, Dean Ing, Jerry, and myself who were also scientists and engineers. The authors became an important element of the Council because it was essential to produce white papers that were readable and interesting. The writers were wordsmiths who could make scientific and technical subjects understandable to nontechnical readers. The Council members boasted an incredible level of expertise in a broad number of technological, political, financial, economic, scientific, military, and business areas.

Each individual Council member represented only himself or herself. No affiliations would ever be attributed. Everyone spoke as individuals and signed off on the white papers only with their names and areas of expertise. No one was ever associated with any company or government organization.

The Council white papers were consensus documents. However, no Council member had to sign anything with which he/she disagreed. Over the years, the various Council white papers represented some of the best untrammeled aerospace, technical, military, and commercial thinking of the time. They were also shining examples of readable documents because of the input of the published authors who took "aerospacespeak" and translated it into readable, understandable everyday English.

The first meeting of the Council was held over the weekend of January 30 to February 1, 1981, in the home of author Larry Niven in the Los Angeles area. Pournelle assumed the task of Council chairman because he's a natural leader. Furthermore, in such a highly vocal group, he could always speak more loudly than anyone else because he'd commanded an artillery battalion in Korea. We would address a problem or policy matter, decide the general position we wanted to take, then break into working groups to tackle smaller elements. Recommendations hammered out during such Council meetings were then posted on the BIX on-line computer network and further massaged in the weeks following the meeting. The resulting Council white papers were indeed read in the White House.

Dr. Hans Mark, then Deputy Administrator of NASA, came to the Council in 1984 and asked us to support Space Station Freedom. The Council looked at the entire program very thoroughly. With the information available at the time, we produced a supportive white paper that, had its recommendations been followed, would indeed have resulted in a space station by 1990.

Then the space shuttle Challenger blew up in January 1986.

The Council met in Niven's home on May 9–11, 1986. The result was an 84-page white paper entitled, "America: A Spacefaring Nation Again." Henry Vanderbilt and I edited the booklet that was published and widely distributed by the L5 Society in Tucson, Arizona.

Among the basic points and recommendations of this report were:

(1)   the national space program is dying,

(2)   the space shuttle is wearing out,

(3)   NASA and other governmental space organizations are not up to the challenge of change, are poorly managed, and are inefficient,

(4)   current launch systems are too expensive and it takes too long to get something into space,

(5)   private enterprise should run space transportation,

(6)   the NASA space shuttle must be flown immediately with military crews if necessary,

(7)   the government monopoly on space transportation must be phased out,

(8)   the federal government should create incentives and adopt risk reduction programs to encourage private space transportation,

(9)   new space transportation vehicles must be developed using current technology and designed for low-cost operation, and

(10) we, as a nation, must stop studying solutions, start flying experimental vehicles within two years, then build and fly spaceships.

Others such as Richard Feynman and Charles Yeager were saying the same things.

The Council reports were read but nothing happened because the entire nation was in shock as a result of the failure of Mission 51-L and the loss of seven astronauts including a school teacher.

In 1988, the Council met twice again. We wanted to find a specific spaceship type that could be suggested for immediate government backing. We'd recommended a new type of spaceship in the 1986 report. However, NASA was playing "business as usual" with the space shuttle and seemed unable to move in any forward direction.

Gary Hudson and Max Hunter are Council members. Hudson had briefed the Council in 1982 about SSTO, but the Council did nothing because, save for some disquieting misgivings about the space shuttle, none of us believed that a new spaceship concept could be sold at that time. Hudson was still ahead of his time, but not by much.

At the first 1988 Council meeting, Hunter presented a briefing to the Council on an SSTO concept called SSX for SpaceShip Experimental.

SSX looked like a Phil Bono spaceship.

SSX looked like Gary Hudson's Phoenix SSTO.

SSX looked like Max Hunter's X-rocket.

SSX *was* all of the above.

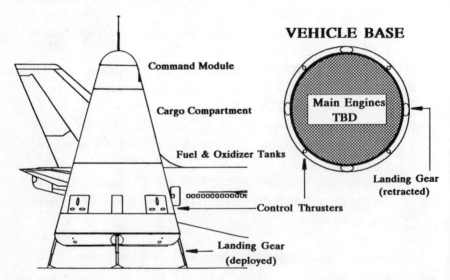

FIGURE 8-4: *The SpaceShip EXperimental (SSX) as proposed by Max Hunter and endorsed by the Citizen's Advisory Council. This is a drawing from the briefing given to Vice President Quayle. The SSX is shown standing next to the tail of a Boeing 747. (Courtesy* The Journal of Practical Applications In Space.*)*

SSX would carry ten tons to LEO. It would be reusable and reflyable within 24 hours after landing. Three experimental prototypes would be built and tested to prove the design. Then it would be produced for commercial sale to the government and to any corporate entity that wanted to build and fly spaceships.

After much discussion and another weekend meeting, the Council members supported this evolution of the SSTOs of Gary Hudson and Max Hunter. A white paper and supporting technical documentation were prepared.

The Council believed that SSX could be built in three years for less than a billion dollars and that it would perform as advertised.

On February 15, 1989, three Council members—Dr. Jerry E. Pournelle, Maxwell W. Hunter, and Lt. Gen. Daniel O. Graham, AUS (Ret)—met in the Executive Office Building of the White House with then-Vice President Dan Quayle.

Graham told Quayle, "I'd like to explain what it is that we are asking you to do. Bear in mind that, as the name implies, we are proposing an *experiment*, not a full-blown program. We believe it

FIGURE 8-5: *Vice President Dan Quayle (left) meets with Daniel O. Graham, Max Hunter, and Dr. Jerry Pournelle on February 15, 1989. This meeting ended the long development of the SSTO on paper and turned it into flying hardware. (Courtesy Daniel O. Graham, Space Transportation Association.)*

tremendously important to find out whether an SSX can be successfully built and flown given today's technology. We think it can, and we stand ready to go into whatever level of technical detail desired by you and your staff. . . . Our proposal runs contrary to the instincts and practices of much of the current space 'establishment' in government and will most certainly generate strong opposition in some quarters of the bureaucracy because it will threaten the status quo. For this reason, this concept must have a champion at the highest level of government . . ."

Vice President Dan Quayle became the first champion of the SSTO. And, over the next five years, the "current space 'establishment' in government" did indeed generate the most powerful opposition imaginable—some of it unlawful, some of it unethical, and most of it unsupportable.

PART II

TESTING
THE CONCEPT

THIS PART OF THE SAGA covers the "present" from February 1989 when Vice President Quayle championed the SSX concept to the end of 1995. During this period, the first reusable rocket was built and flown. A true SSTO was not. One reason was the step-by-step approach that was taken: "Build a little, fly a little." It was not as bold as the approach taken by those who wished to stop the program because Part II involves what has become known as a "paradigm shift," a fundamental change in an accepted manner of thinking. Making the paradigm shift happen was far more difficult than making the off-the-shelf technology work.

This part also may help answer questions that arise in the minds of many people, "Why can't we solve problems quicker? Why do we have to live with What Is when we have a clear vision of What Can Be and know how to get there from here?"

The problem that exists today with space transportation is the fact that the United States won the Moon race then continued to ride a horse that should have been put out to pasture.

However, America is now in another space race that is an economic one. With the end of the Cold War and the emergence of the worldwide market economy, the question is not which ideology or form of government will prove itself superior on the basis of what country does more in space. Instead, it's a matter of who will dominate the commercial space age of the twenty-first century by making money doing it just as Americans have dominated commercial aviation in the twentieth century.

Part I started halfway to nowhere. Part II begins halfway to somewhere.

NINE

SDIO and Phase I

POURNELLE, HUNTER, AND GRAHAM surprised Vice President Quayle by suggesting that NASA should not be given the job of developing the SSTO spaceship. As Graham wrote in his autobiography, *Confessions of a Cold Warrior,* "He was even more puzzled when I opposed the second choice, Air Force. I urged that the job be given to SDIO (the Strategic Defense Initiative Organization) as the organization without enough colonels and ranking civilians to louse up the effort."

A brief explanation of this rationale is in order and should be kept in mind on the basis of later happenings.

At that time, NASA had the reputation for taking a project and turning it into a multi-dirksen megaprogram lasting ten years. The space agency was deeply involved in planning the "next generation launch vehicle family" of expendable rockets. Over the years, this program was called the Advanced Launch System (ALS), then the National Launch System (NLS), and then Enhanced Expendable Launch Vehicles (EELV). The NLS acronym was waggishly corrupted to the "Never Launch System." No hardware was fabricated or tested but many reports and studies were produced. The program managers lost sight of the fact that rockets on paper don't launch payloads into space. NLS also was typical of government projects with regards to timing and costs. NASA proposed launching the first NLS expendable rocket in about 2002 after spending 14.4 dirksens on the program.

Although the SSTO offered a fresh approach to space transportation, few people in the government and the aerospace industry
saw it that way in 1990 because "everyone knew" that an SSTO
couldn't be built. They'd forgotten that engineers had said it could
be done as long ago as 1945.

The United States Air Force was involved in the ALS/NLS
studies because it wanted a new series of high-tech expendable
launch vehicles to place various military space assets in orbit after
the turn of the century. However, Air Force generals said they didn't
have a "mission" for an SSTO and anything that doesn't have a
"mission" doesn't find any support in the Department of Defense (DoD).

The Strategic Defense Initiative Organization (SDIO) indeed did
have a mission for the SSTO. This organization had the task of
developing and deploying a system to protect the United States and
its allies against ballistic missile attacks. Dr. Lowell Wood, Director
of the Lawrence Livermore Laboratories and a member of the
Council, had come up with the concept of "Brilliant Pebbles." One
of the ways to intercept and destroy an incoming ballistic missile
warhead is to hit it hard and fast with a high-tech rock using a smart
guidance system. Wood and his researchers had come up with a
way to miniaturize such a "kinetic kill" weapon so it was no longer
a "smart rock" but a smaller "brilliant pebble." A launch vehicle
was needed for Brilliant Pebbles.

Thus, the SSTO program was given to General George Monahan,
then in charge of the Strategic Defense Intiative Organization, who
assigned it to his Deputy for Technology, Lt. Col. Simon P. "Pete"
Worden who set up an SSTO program office manned by only two
USAF officers, Lt. Col. Pat Ladner and Maj. Jess M. Sponable.

Ladner and Sponable created some new ground rules to define
size as well as ensure they would get aircraftlike operations, quick
turnaround, reusability, and small operating crews that the Council
had told Quayle were possible. These included a fully reusable
operational spaceship capable of carrying a 20,000-pound payload
to orbit. It should have a seven-day turnaround between flights with
no more than 350 maintenance man-days per flight. The design
should include a safe engine-out and intact-abort ability like an airplane. It should cost less than $5 million dollars per flight and have

a 99.5% or better flight reliability. It should be capable of being flown manned or unmanned from austere launch facilities with a small ground crew. Some of the SSX concept remained in the SSTO program, but the vehicle had grown bigger and heavier. Hunter and the Council members fought this at the start. Some people don't understand that you have to build a DC-3 before you build a DC-10.

Ladner retired in 1991 before the program really got moving, but Sponable stuck with it.

Jess Mitchell Sponable is a new-generation space pioneer. He graduated from the United States Air Force Academy in 1978. He was selected for the second group of Department of Defense manned spaceflight engineers in January 1983 but left in 1985 without flying a space shuttle mission. He went to the Air Force Space Division in Los Angeles and then served as project officer for the X-30 NASP program where I first met and worked with him in 1988.

Ladner and Sponable established a three-phase program.

Phase I was a $15-million competitive concept definition program to review existing SSTO concepts, to refine the concepts, and thus to reduce technical risk.

Phase II would involve building and flying an experimental sub-scale vehicle whose purpose was to test and prove the unknowns in the SSTO concept. The philosophy was "build a little, fly a little," not produce paper reports and overhead transparencies.

Instructions from Ladner and Sponable for both Phase I and Phase II said that the experimental suborbital test vehicle must be built with off-the-shelf hardware. There would be zero dollars for research and development. This was unheard-of in the aerospace community and caused some real heartburn in some aerospace management conference rooms and design meetings.

Phase III would be the development of a vehicle capable of attaining orbit. But it would *not* be a prototype. It would still be an experimental vehicle. Its design would incorporate what was learned in Phase II by actually flying a reusable rocket. In short, the government would not own the design as NASA owns the space shuttle. The SSTO design and technology would remain the property of the developing company who would be able to build and sell spaceships to operating companies and to the government.

The basic development strategy of the SSTO was to demonstrate the technical feasibility of SSTO spaceships and thus reduce the perceived technical risk that scares away private financing. However, a continual battle raged to keep SSTO from becoming Super Shuttle, Shuttle II, or "Son of Shuttle."

Please keep the three-phased nature of the SSTO program in mind. We will encounter it again.

On a competitive basis, four aerospace companies were selected to look at different approaches to SSTO in Phase I.

Compiling the Phase I history of SSTO has been a historian's nightmare. Most of the documentation that survives is in the form of poor photocopies of the standard aerospace industry briefing tool, the overhead transparency. The written reports and proposals are usually tossed out because aerospace managers have a poor sense of history. In addition, they need the file cabinet space for the next project. Aerospace engineers will often save little personal mementos. However, they will trash important historical documents if the corporate historian doesn't stay on the ball and rescue these before they end up in the dumpster.

Therefore, because the losers of Phase I didn't bother to save much, few illustrations of the proposed SSTO concepts of those studies survived. I was moderately successful in obtaining documents from General Dynamics and McDonnell Douglas at the time, but I could get nothing from Boeing or Rockwell International. Therefore, the illustrations of the Phase I SSTO concepts were what I could get and it's a miracle any survived at all.

General Dynamics received a contract to study a Vertical-Take-Off-Vertical-Landing (VTOVL) rocket with a plug nozzle propulsion system and base-first entry. They called it the Millennium Express. The sub-scale Phase II experimental vehicle was called the Pathfinder. I have excellent documentation, including the final proposal along with drawings because Daniel Heald, Tom Kessler, and Rick Jurmain of the proposal team asked my son to build some inexpensive desk models for them.

McDonnell Douglas proposed a VTOVL rocket with a plug nozzle but using nose- or side-first entry like a ballistic missile warhead. One of the big problems with this spaceship, called the Delta

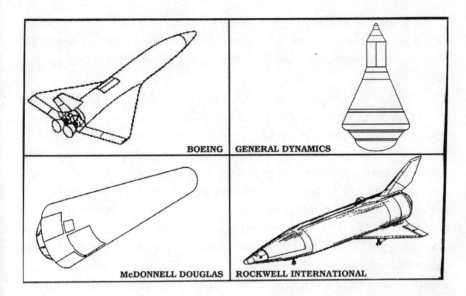

FIGURE 9-1: *Rough drawings of the four contenders in the SDIO Single-Stage-To-Orbit Phase I competition. (Courtesy BMDO SSRT program office.)*

Clipper, was its marginal stability during some entry conditions. McDonnell Douglas engineers subsequently solved this problem by proper combinations of shaping, maneuvering flaps, and small thrusters. The original Delta Clipper DC-Y proposal resembled an upside-down ice cream cone.

Rockwell International proposed a Vertical-TakeOff-Horizontal-Landing (VTOHL) ship that resembled the space shuttle orbiter and, like the shuttle, used wings to land on a runway. Rockwell obviously relied heavily on their design, construction, and operational experience with the 20-year-old space shuttle orbiter technology.

Boeing came in with a Horizontal-TakeOff-Horizontal-Landing (HTOHL) concept that was basically an updated rehash of some of their earlier work and required either a ground launch from a big rocket sled or an air launch from the top of a modified Boeing 747 airliner. Thus, the Boeing bid wasn't really an SSTO.

Another proposal in Phase I came from Leo Cormier of Third Millennium, Inc. who again put forth his Space Van concept, a winged orbiter launched at high altitude from a Boeing 747 like Boeing's entry. Third Millennium didn't survive the Phase I cut.

Also left of out the Phase I studies was Grumman. I've been consistently unable to get any information on their Phase I proposal because the company was basically going down the tubes at that time.

Strangely, in light of later developments, Lockheed was asked to bid Phase I and declined to do so. This could well have been because of the paucity of funds in the program. In Phase I, $15 million were to be divided among four contractor studies. The Phase II contract was to be $60 million. This is not enough money for some aerospace company executives to get excited about. As a result, it appears that some firms got involved only because a group of SSTO enthusiasts in the company wanted to try getting the contract. However, they were never supported by the next management layer.

This was true in the case of General Dynamics whose Phase I team of Heald, Kessler, Jurmain, and others later became quite vocal about the lack of corporate support and the failure of vision of higher-ups, nearly all of whom were Atlas ballistic missile types and not airliner or even fighter plane engineers. On March 28, 1991, in the middle of the proposal effort, the GD SSTO proposal team got a new division general manager. "He knew it couldn't be done and said so. Never mind the detailed analysis; he knew the *truth*. As for the plan to build and fly something in a year, he considered it ludicrous and said he would fire anyone who broached the subject with him again," was the personal message I got from Jurmain.

Of course, at that time in mid-1991, the general belief was, "SSTO will never work. Why waste our time with it?" The people at NASA believed this and continually said so publicly. Some of the aerospace company executives, never ones to stray far from the guiding golden words of their primary customer, echoed these beliefs. Fortunately, in hindsight, this was a blessing because it allowed Phase I and Phase II of the SSTO program to move ahead without any opposition. Since it "couldn't be done," no one paid any attention to it because they "knew" it was going to fail. This helped keep it low-profile. The trouble came later when it worked and became high-profile.

I heard frank admissions from some aerospace executives that they'd allowed their engineering teams to bid on the SSTO Phase I contract only because there might be something there and they didn't want to be left out if there was. They hedged their bets but were generally unwilling to put internal research and development money into the proposals.

Some proposers apparently didn't believe Ladner and Sponable when it came to the "off-the-shelf" requirement. In the past, it had always been possible to justify a program over-run of man-hours or funds because some particular technical goodie looked too promising to pass up and including it would have made the product so much more efficient, powerful, fast, etc. Usually, the NASA or USAF program managers, civilian or military, went along with this.

Not this time. Ladner and Sponable monitored the proposal development as closely as possible and axed most of this proposed research and development activity.

For example, some of the proposals included exotic ideas that were quickly weeded-out on the basis of the "off-the-shelf" requirement for the Phase II rocket. Among these was a propulsion scheme that envisioned using the super-cold liquid hydrogen in the rocket to liquefy surrounding air during ascent, thus creating the rocket's oxidizer supply. This is known as a Liquid Air Cycle Engine (LACE). No dice, said Ladner and Sponable. That was "vaporware," not hardware. It wasn't sitting on the shelf waiting to be used. Forget it.

However, the winning concept did use some technology that wasn't off the shelf. But this was quickly rectified when it was made clear by Ladner and Sponable that the proposal had won only on the basis of overall technical merit. Once the contractor had been selected, the company was instructed to modify the design proposal to conform to the "off-the-shelf" requirement.

Finally in the summer of 1991, all the study proposals were turned in to the little cluster of rooms crammed into the basement of the Pentagon where Ladner, Sponable, and a couple of secretaries were *the* SSTO project office. Then the two officers carefully studied each proposal.

On August 16, 1991, they awarded McDonnell Douglas Space Systems Company a $58.9 million contract to design, build, and flight test the one-third-scale Delta Clipper Experimental (DC-X) reusable rocket within two years.

It looked like the end of decades of aborted paper studies and the start of real SSTO hardware tests.

The Delta Clipper

McDonnell Douglas Space Systems—identified hereafter by the acronym MDC for McDonnell Douglas Corporation although it was their Space Systems Company that got the contract—wasted no time putting together their team. They didn't have much time in the first place. The contract required them to build and fly a one-third scale prototype, the Delta Clipper Experimental or DC-X, within 24 months. In the aerospace business, this was an impossibly short time to design, build, and test a rocket. Max Hunter had done Thor at Douglas in less than a year, but that was a third of a century before 1991, and things had not changed for the better.

On September 5, 1991, I received a package from Dr. William A. Gaubatz, Director of SSTO Programs at MDC. I'd worked with Gaubatz when I was a consultant to MDC on the X-30 NASP program. In the package was a set of overhead transparencies showing the new Delta Clipper design.

It was called Delta Clipper in honor of the company's successful Delta rocket (a grown-up version of Hunter's Thor) and because it was supposed to be like the 19th-century commercial clipper ships.

The Delta Clipper had changed in the few months since I'd received the brochure prepared for the contract briefing. It had morphed into a flared conical shape with progressively flattened sides. It now resembled a reentry lifting body, a wingless vehicle that, at supersonic and hypersonic speeds, actually would fly like a winged vehicle. If you take a delta-wing airplane and squish it here and fill it out there, it can be morphed into such a shape very easily.

The shape came from thousands of hours of wind tunnel testing performed years before to find the best shape for the maneuverable independently-targeted reentry vehicles (MIRV) used on Minuteman ballistic missiles.

However, the Delta Clipper concept still used the segmented plug nozzle rocket engine.

Along with several Council members—Jerry Pournelle, Larry Niven, Lowell Wood, Max Hunter, and Steve Hoeser—I received an invitation to attend the SSTO Phase II Kickoff Meeting at MDC in Huntington Beach, California, on October 29, 1991. Pournelle, Niven, Hunter, and I had dinner the night before the meeting to discuss the Delta Clipper that had emerged from the Council's SSX proposal to Quayle.

The Council hadn't proposed a sub-scale subsonic experimental vehicle that Pournelle tagged as "single-stage-to-twenty-thousand-feet." The Council had been bold in its proposal, believing that going straight for orbit with a series of full-scale experimental vehicles was the way to do it. This didn't mean that we wanted the SSX to try for orbit on the first flight but to start low and slow, eventually "pushing the envelope" higher and faster until the SSX went once around and landed. But Ladner and Sponable, operating on more conservative principles, opted for the sub-scale DC-X. Okay, we figured something could be learned with the DC-X if they didn't get hung up on the idea of testing it forever.

The orbital experimental version of the Delta Clipper, tagged DC-Y, had grown. It was nearly double the lift off weight of the SSX recommended by the Council and briefed to Quayle. It seemed to us that MDC was aiming to build a space DC-10 when what the Council had determined to be feasible and needed was a space DC-3. Hunter pointed out that the old paradigm of "bigger is better" no longer held, and we discussed the marketing and economics of trying to fill a commercial DC-Y derivative to the break-even load factor versus doing it for the smaller SSX. We decided that Gaubatz should be told to "think small."

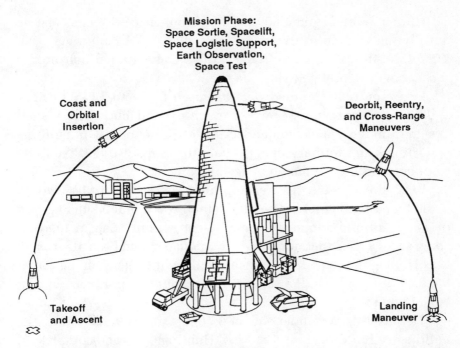

Mission Phase:
Space Sortie, Spacelift,
Space Logistic Support,
Earth Observation,
Space Test

Coast and
Orbital
Insertion

Deorbit, Reentry,
and Cross-Range
Maneuvers

Takeoff
and Ascent

Landing
Maneuver

FIGURE 10-1: *By the time of the kickoff meeting in October 1991, the configuration of the Delta Clipper had been modified with flat sides. This drawing also shows how the Delta Clipper would be operated. (From the McDonnell Douglas meeting brochure.)*

In addition, we knew that the plug nozzle rocket engine, efficient and neat as it was for this application, would have to go. The reason was simple: time. The DC-Y was targeted to be flown to orbit by 1996, the original goal for the SSX that the Council believed possible in 1989. And in late 1991, MDC showed a chart indicating a first orbital flight in mid-1996. *In light of what was to happen, this mid-1996 orbital schedule should be remembered.*

In 1991, no one had an operating plug nozzle rocket engine with a thrust of a million pounds. Rockwell International and Aerojet General had both conducted experimental firings of plug nozzle and aerospike rocket engines in the 1960s and the early 1970s, but we didn't know about it. In the intervening years, the concept had fallen into limbo because there was no application for it and no available government development money.

Hunter pointed out that the development of a rocket engine usually takes three to five years, regardless of the funding level. Therefore, the DC-Y would have to use conventional bell nozzle rocket engines, preferably off the shelf.

Jerry Pournelle pointed out that the DC-X and DC-Y designs we'd seen before showed far too much expendable artillery shell design thinking. He recommended that MDC's expertise in building jet airliners and military fighter planes be tapped so that the Delta Clippers wouldn't be maintenance nightmares.

Why did we have the presumption to believe that MDC would listen to us? In the first place, the Council was a independent group of cognizant and supportive people just as interested in the outcome as MDC. Furthermore, Ladner, Sponable, and Gaubatz trusted Max Hunter and the other engineers and technologists on the Council. Last but not least, Pournelle and Graham had political clout in Washington.

The meeting got under way at 8:00 A.M. in the large, futuristic, multi-story Building 14 in the MDC Huntington Beach complex on Bolsa Avenue. A total of 88 people, including five people from NASA, found seats in the room. Arranged around on tables were examples of DC-X components—a Pratt & Whitney RL-10 rocket engine, various hydraulic actuators from the Delta rocket, an autopilot from the McDonnell Douglas MD-11 transport, etc.

What we saw and heard that day were the results of only about 60 days of intense team building and design work that had commenced upon signing the contract. The Delta Clipper Team didn't know a great deal about the DC-X yet although they were pushing hard because they had to build fast and fly soon. The DC-Y follow-on was still "vaporware," in contrast to hardware.

One aspect of this bears telling. The Delta Clipper Team was about 100 top MDC engineers. Their goal was unusual: A reusable rocket didn't exist, and they had to design, build, and operate a proof-of-principle prototype. This would be a cheap-and-dirty off-performance piece of iron mongery using off-the-shelf technology and hardware, the DC-X. It would be used to check out some of the questionable approaches to the solution, find out if the approach was really workable, discover the items that are always overlooked

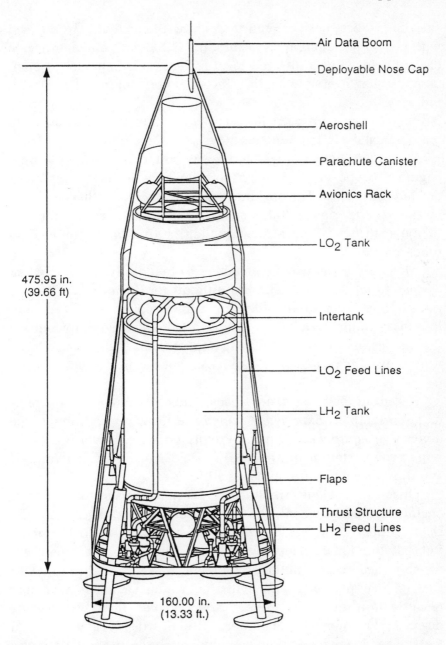

Air Data Boom

Deployable Nose Cap

Aeroshell

Parachute Canister

Avionics Rack

LO$_2$ Tank

Intertank

LO$_2$ Feed Lines

LH$_2$ Tank

Flaps

Thrust Structure

LH$_2$ Feed Lines

475.95 in.
(39.66 ft)

160.00 in.
(13.33 ft.)

FIGURE 10-2: *Cutaway drawing of the Delta Clipper DC-X, the one-third scale experimental vehicle intended to test various SSTO concepts. (From McDonnell Douglas DC-X press kit.)*

even in the best designs, and then allow the company to proceed with the pre-production spaceship, the DC-Y, with a higher degree of confidence and a lowered level of risk. This approach is standard, old-hat, everyday engineering anywhere but in the aerospace industry.

But, in the aerospace industry, no one had been allowed to make a mistake in the previous 30 years.

It was fascinating to watch the way both experienced old-time engineers (who are now managers) and fresh-caught engineers tackled the project. The old-time engineers had to battle decades of on-the-job experience. Tattooed on their brains was the dictum: "Thou shalt not fail, it must work the first time, and thou hast no room for error."

These older engineers remembered, often dimly, the time when it wasn't that way. It wasn't too difficult for them to shift mental gears and get back to the old method that amounts to: "Well, hell, let's just whomp up a boilerplate test model of this puppy and see if it passes the smoke test when we plug it in!" That's what engineering used to be all about, and it's one of the elements that made it fun.

Except in the modern aerospace industry, engineering operates on the principle, *"Experience gained is directly proportional to the amount of equipment ruined."* Prototypes weren't worth a damn unless you busted them. Otherwise, you'd underestimated yourself and didn't need the prototypes at all.

Once that ancient principle was reestablished in the minds of engineering management, the project actually became fun for the Delta Clipper Team and contributed to its incredibly high morale. But that didn't make it any less stressful. The lack of big money and the short deadlines kept the pressure on.

The real problem came with the young engineers who had recently (within the last ten years) received their engineering degrees. The young engineers were brilliant when it came to design work. They knew how to run computer analyses until the floor was covered with printouts. They were whizzes with Computer Aided Design (CAD). But they'd never "bent tin." They'd never been responsible for designing something that could be built and was supposed to do something. This puzzled me at first. Then I figured out what had happened.

In 1978, my son decided he wanted to be an engineer so he could become a product designer. We went to several colleges and universities to investigate their engineering curricula, facilities, and teaching staff. I learned something had changed in Engine School. Two career paths existed (and still exist) for engineers. An engineering degree now consists of an extremely strong emphasis on scientific theory, mathematics, and computer technology. *And practically no hands-on lab work!* The venerated engineering degree had been converted into a degree in applied science!

The other path led to a bachelor's degree in "engineering technology." Upon close investigation, I discovered that this poor stepchild of modern undergraduate study was indeed the sort of hands-on, practical engineering curriculum that I was familiar with back at mid-century. But it no longer turned out "engineers." It graduated lowly "engineering technologists."

No wonder some modern products seemed to be less-than-elegant in their design, construction, and operation! Essentially, they're designed by scientists, not engineers! The real grubby-handed engineers, now called "engineering technologists," take over after the "engineers" are finished. They're the ones who have to make the product work after the design has been approved!

By the way, I have nothing against scientists. In fact, my degree is in physics, not engineering. Scientists are needed to explain why something works after the inventor conceives it and the engineers make it work. Yes, a few modern products have sprung from the science lab. But far more of them have come from inventors.

The Delta Clipper Team learned the hard way what engineers used to learn in undergraduate work and their first few years in the field. They had to bend tin against a schedule. They had to make do with what they could get off the shelf. They didn't have one thin dime for research and development. They learned to read Thomas' *Register.* They learned to scrounge through junkyards to find something cheap that would do the job. They faced a world where Good Enough is the enemy of the Best, where an adequate solution today is far more important than a perfect solution tomorrow.

In the long run, the Delta Clipper Team performed in an exemplary fashion. In fact, I could see the difference when I was invited to the follow-on User's Meeting on March 10, 1992. It was obvious that MDC had indeed listened to comments and suggestions made in the October 1991 meeting. The DC-X design was frozen at that point, of course, but the follow-on DC-Y had been carefully studied and refined.

Gone was the plug nozzle rocket engine. In its place was a cluster of eight bell nozzle rocket engines, four with extendable nozzle skirts to take care of the altitude compensation problem.

The DC-X design and the DC-Y concept showed the results of adding to the Delta Clipper Team some engineers from McDonnell Aircraft Company in Saint Louis and from the Douglas Aircraft Company in Long Beach. The Saint Louis contingent had designed such aircraft as the F/A-18 Hornet and the AV-8B Harrier. The Long Beach group came from the MD-80 and MD-11 jet airliner programs.

The DC-X was in the tin-bending stage. The DC-Y was in preliminary design. In short, every nut and bolt of the DC-Y hadn't been specified yet. But it was still a monster designed to loft 24,900 pounds to LEO with a dry weight of 89,000 pounds, a rocket propellant weight of 1,147,600 pounds, and a gross liftoff weight (GLOW) of 1,279,000 pounds. This included a design margin of 15%, which is the number that engineers figure in to account for the usual optimistic weight determination. The DC-Y would be 127 feet high. As time went by, the GLOW crept up to 1,400,000 pounds and the payload increased to 40,000 pounds.

We Council members kept insisting that the DC-Y should be sized for no more than 20,000 pounds to orbit. This was "medium lift" capability that would handle a very large percentage of the satellite lift market. Independent studies showed that even 10,000 pounds of payload would be economical at first. But the design payload kept right on growing as engineers said, "Hey, we can boost another five thousand pounds if we just make this little improvement!"

The big problem was (and still is) the rocket engines. Multiple rocket engines were necessary because of the requirement for safe abort and landing. An "engine-out" safe abort capability is a non-negotiable requirement if a rocket is to operate like an airplane that *does* have an engine-out capability. All of the Saturn rockets had it and even the space shuttle has it to some extent.

A rocket engine producing about 250,000 pounds of thrust is needed. It should use liquid oxygen and liquid hydrogen. Other propellants were later suggested. It must be capable of being throttled from 30% to about 110% of operating thrust with normal flight thrust set at 90%. This concept is no different than designing an aircraft engine that is de-rated or operated at less than its maximum output.

The engine must use a safe operating engine cycle, and the safest one of all is the one used by the Pratt & Whitney RL-10 engines. This is the "expander cycle." The Space Shuttle Main Engine doesn't use this cycle, and its turbopumps have been a problem since Day One. The shuttle engine is also maintenance intensive, to put it mildly. After each shuttle flight, the engines are de-certified for flight use, shipped to Canoga Park, California, for a complete rebuilding, and then completely retested for about 80% of the individual engine's design life. The next time you see a space shuttle launched, think of it as being powered by well-used engines nearing the end of their useful lives.

It's of interest here to point out that no new rocket engines have been developed in the United States for more than 25 years. The space shuttle engine was the last one.

The engineers at Pratt & Whitney said they could build a larger version of the reliable RL-10. But the price and the time requirements were unsuitable. Delta Clipper needed a modification of an existing rocket engine.

However, both Aerojet General and Rocketdyne once built rocket engines that would fill the bill. For example, the Rocketdyne J-2 was used in the Saturn moon rockets. It has been modified and tested in throttleable condition. It has the efficiency (specific impulse) that will allow the DC-Y to be orbital with a payload. It's being ignored by engineers who are infatuated with what they've learned about Russian rocket engine technology. Or the engine-makers are waiting for several million dollars from NASA to develop the perfect SSTO engine. So the old Cold War missile paradigm remains to be totally smashed in this regard. It will be.

FIGURE 10-3: *One of the four Pratt & Whitney RL-10A-5 rocket engines used to propel the DC-X. (Pratt & Whitney photo from DC-X press kit.)*

Even in 1991, we knew the SSTO had an engine availability problem. But it was and is a problem that can be solved from an engineering or technical point of view. Indeed, off-the-shelf engine designs could be used for the early SSTOs. Although these off-the-shelf rocket engines aren't optimized, *they will work.* As better engines become available, SSTOs can be retro-fitted or redesigned to accommodate them.

However, while SSTO enthusiasts and engineers were wrestling with these erudite technology problems, the Delta Clipper Team was bending tin on the DC-X against a deadline. Few of us realized that once we'd "started this riot," as Max Hunter put it, it would quickly become something none of us had anticipated. None of us foresaw that it would flare to the intensity it did.

As we have seen thus far, the SSTO problem is *not* technical. We have the technology. It is *attitudinal.* It represents a confrontation between a now-possible way of doing something better versus an antiquated way of thinking that was useful in its time but whose time has passed. The old saying, "There is nothing as powerful as an idea whose time has come," must be modified to, *"There is nothing so impotent as an idea whose time has passed."*

ELEVEN

Roll Over and Roll-out

THE SSTO PROJECT in the Strategic Defense Initiative Organization (SDIO) actually was stillborn. In 1991, Ambassador Henry Cooper took over as director. Late that year, he was reminded by the Office of the Secretary of Defense that his job was to figure out how to shoot down incoming ballistic missile warheads and not to develop space launch vehicles. He was told that new space vehicle development, if done at all in the Department of Defense, had to be shared with NASA.

We think we know who put this restriction on Ambassador Cooper. It was one or two staff people in the Pentagon and Congress who were waiting for the proper moment to destroy the SSTO project. Being civil servants and not political appointees, they couldn't be fired. And, having stayed on the job for many years, they'd accumulated a lot of information about skeleton-filled closets and buried bodies inside the Beltway. Judging from what happened in the 24 months following the award of the Phase II contract, it's now apparent that the vested interests who would arise eventually to oppose the SSTO were (1) keeping quiet for the moment, hoping the DC-X would fail and (2) setting the necessary political traps just in case the DC-X *did* work.

The SSTO advocates anticipated this. Graham had warned Quayle of it during the February 1989 briefing.

However, the SSTO advocates did *not* anticipate the intensity and viciousness of what was to come. In some ways, it was impossible to prepare all possible defenses against attacks because so many avenues of assault are available in Washington. Apparently only Vice President Dan Quayle's role as champion of the SSTO kept the program from even harsher attacks while it was still in Phase I.

Then, in the 1992 election, SSTO lost its champion. Without Vice President Quayle to protect the program, its opponents moved to stop Phase II in such a way that they couldn't be blamed for it.

The first indication that pressures were being brought to bear was the change of the project's name. It became the Single Stage Rocket Technology (SSRT) program. In short, the project shifted from one charged with eventually building an orbital SSTO demonstrator vehicle to a "technology development" program.

This move has killed every new launch vehicle program since the advent of the space shuttle.

The Council members saw the potential consequences, but erroneously believed that getting the DC-X flying was the most important step in the program. The mistake was focusing on the technology, but the technology wasn't the show-stopper.

SDIO also went through a name change and became the Ballistic Missile Defense Organization (BMDO).

Not all of this change was caused by Dan Quayle's failure to win reelection. Part of it might have been due to the fact that BMDO's chief engineer, Mike Griffen, a staunch SSTO advocate, left to become head of NASA's Office of Exploration, a post that was created for him. Congress axed that slot a year later. Griffen then became NASA "Chief Engineer," another post created for him, more or less as a way to keep him happy by giving him a desk and filing cabinet but no secretary or telephone. I'm not sure if Griffen was co-opted to NASA to remove his influence in the Ballistic Missile Defense Organization, but certainly one of the best ways to disarm the opposition is to shift the leaders and strong advocates into positions of zero power.

Then Col. Pat Ladner retired, and Col. Simon P. "Pete" Worden bucked up Major Jess M. Sponable to Ladner's position as project officer.

In a strange move, Worden then killed the DC-Y. All of the $30 million authorized and appropriated to start design work on the orbital DC-X follow-on was reprogrammed to the DC-X portion of the project. Perhaps Worden had to do it because $30 million wasn't enough to complete the DC-X Phase II project and not enough to do much more on the DC-Y.

Unfortunately, the $59 million of Phase II money turned out to be barely enough to build and test only one DC-X. In this regard, Worden was probably correct in reallocating the DC-Y development funds to the DC-X construction and flight test program.

Building only one experimental test vehicle is indeed unusual in an X-vehicle program. Three X-15 rocket planes were built and flown by NASA and USAF between 1958 and 1967. Even the experimental XF-22 and XF-23 advanced tactical fighter program had several airplanes available for flight testing. The reason for multiple experimental vehicles seems simple enough: One of them is certain to be "bent" or "pranged" in the course of the test program as the design is pushed to its limits. Without a back-up, a single vehicle program thus grinds to a halt while money is found to build another one. This didn't happen in the X-15, XF-22, and XF-23 programs because they had multiple test vehicles from the start. As any real engineer knows, it doesn't take twice as long or cost twice as much to build two of something while you're at it.

And—although Sponable and Gaubatz didn't like to hear it—they were reminded by several of us on the Council that if they didn't bend the DC-X, they probably didn't need it in the first place. However, they felt secure in the fact that they had a large number of spare parts on hand. They believed they could fix the DC-X if it didn't get bent too badly during flight tests. It turned out, of course, that a second DC-X was really needed after all.

But a back-up to DC-X was not to be. The intent on the part of BMDO was to keep the Phase II SSTO program low-budget and low-profile to reduce the perceived threat to the Pentagon/NASA "Gun Club" of expendable launch vehicle people as long as possible.

They couldn't do it.

The Gun Club found out anyway because the project's profile wasn't low enough. They realized the threat to their expendable launch vehicle bureaucratic and industrial empires was real if DC-X turned out to be successful. In the meantime, they floated the usual "expert opinions" that SSTO was "impossible" and started another campaign saying that the DC-X was a "stunt" and wouldn't prove anything.

Many of us who watched this take place were reasonably sure that the members of the Gun Club were starting their campaign to kill SSTO.

If the DC-X didn't fail technically then the Gun Club wanted to make sure that it failed politically.

By January 1992 it became apparent that SSTO was facing a curious dilemma: It had attracted too much support! The National Space Society and John Pike of the Union of Concerned Scientists, organizations that hadn't been supporters, came out cheering for SSTO.

SSTO was kept low-profile for several other reasons. BMDO was being pressured by the Air Force and NASA to support heavy lift launch vehicles but didn't need them. Ambassador Cooper's people wanted cheap medium to light lifters for Brilliant Pebbles and didn't desire to tie the Brilliant Pebbles deployment schedules to a launcher research and development program. This weakened the link between SSTO and operational necessity.

Furthermore, BMDO didn't want the interagency and interdepartment grief that would come when SSTO was finally perceived as the Great White Hope of space transportation—which it is—and everyone wanted on the band wagon. So they tried to maintain a low profile by placing emphasis on DC-X as a potential "sounding rocket" prototype rather than an experimental space launch vehicle.

To add to the "riot" that Max Hunter had predicted and in the face of continued NASA insistence that SSTO wasn't feasible, a new player joined the fracas in early 1993. David Urie, head of the Lockheed Advanced Development Project Division—otherwise known as the "Skunk Works" where the U-2 and SR-71 spy planes had been developed and the F-117 Stealth Fighter had been born—began giving "confidential" briefings around Washington on Lockheed's SSTO concept!

Recall that Lockheed had "declined"—I was told that the word "refused" would describe it better—to participate in the SSTO Phase I concept definition.

People who were given these briefings were asked to maintain confidence because the Lockheed SSTO was a privately-funded in-house effort.

However, because Lockheed gave these briefings to a wide variety of people in many organizations, the details leaked. This is a way of life in the nation's capital where the government runs on information leaks. Some information is deliberately given to "calibrated leaks"—i.e., people who will leak just the right amount of information in just the right places in order to achieve the desired results. For example, one of the primary aerospace business publications, *Aviation Week & Space Technology,* is widely known as "Av Leak." Some of us outside the Beltway knew within 24 hours about these briefings and managed to get most of the basic details.

Lockheed presented a Vertical-TakeOff-Horizontal-Landing (VTOHL) SSTO capable of carrying a payload of 40,000 pounds to orbit. It had a strange configuration that looked like a delta-winged airplane. In this regard, it resembled the earlier Lockheed Starclipper design except it didn't drop any tanks on the way to orbit. Its rocket engine was a "linear" design that has its roots in work done by Rocketdyne in the 1970s and then by Aerojet for the NASP X-30 program in the 1980s. A "linear" rocket engine can be described as a plug nozzle rocket engine with an infinite radius.

Some of us wondered what Lockheed was up to. Was this an end-run around the SSTO program? Or was it a trial balloon sent up to test the potential military market for such a vehicle? It's interesting to note that a major aerospace company—one with a long history of innovative airplanes and a Skunk Works that accomplishes the impossible—was then convinced it could build an SSTO and sent its top people to Washington to give briefings. In this regard, Lockheed management did a complete turnabout from the time several years earlier when they'd rejected Max Hunter's X-rocket proposal.

If Lockheed was looking for support from the Air Force for the vehicle, they didn't find it. If they were looking for funding, they didn't find that, either. Unless, of course, the Lockheed SSTO was either an outgrowth of the enigmatic Aurora program or a coverup in the way of misdirection. Aurora may have been the overall code name for a series of Department of Defense technology development programs that included Science Dawn, Science Realm, and Have Region. All of these, now generally unclassified, demonstrated numerous breakthroughs in materials research and manufacturing that benefit the SSTO concept. The dart-shaped aircraft seen over Nevada and the North Sea may have been Mach-8 high-altitude aircraft. They were certainly not spaceships.

This didn't seem to have much effect on SSTO except to compound its problems caused by a low budget. This marginal budget has had two consequences. One of them is good news, and the other one is bad news.

The good news was that it caused project managers, officers, and engineers to think instead of playing the old aerospace game of "spend another million dollars and hire another acre of engineers."

The bad news was that the lack of money continually threatened to stall the SSTO program as its momentum built up in 1993. Things such as trucking bills and cryogenic liquids cannot be "costed-out" of the program by passing government chits around but must be paid for by Treasury checks.

But the Delta Clipper Team, both in the tiny Pentagon basement cubbyhole and at Huntington Beach, consistently refused to roll over and play dead. They had a goal. This was perhaps best summarized by one of the overhead transparencies used by Bill Gaubatz entitled "The Delta Clipper Team Vision:"

> *Successful completion of the SSTO program ensures reliable space access for peaceful use as well as for the defense of free nations, and it establishes a new space highway through which space commerce will flourish. Hardware flying fulfills the promise of our vision.*

One of the overhead transparencies flashed on the screen by Gaubatz during his briefings was a quote from A.C. "Mike" Markkula, Jr., one of the founders of Apple Computer Company:

Some things have to be believed to be seen.

The DC-X showed this to be true when it was revealed to the public on April 3, 1993.

It was a "roll-out" in the grandest tradition of the airliner industry.

It was also the first roll-out of a wingless rocket-powered vehicle. (The North American X-15 was rolled out in 1958, but it was a winged rocket-powered airplane.)

The ceremony took place at the McDonnell Douglas Space Systems facility on Bolsa Avenue in Huntington Beach, California, on the north side of Building 45B, a tall steel hangar. A gaggle of VIPs was present, but most of the crowd of about 300 people were the wives and children of MDC employees. The VIPs were there for the usual reasons. The MDC families were there so they could see what had been occupying the days, nights, and weekends of those family members who'd spent the previous 20 months working on the DC-X.

The building of the DC-X was an unusual effort. A flyable rocket vehicle—*operating hardware*—had been designed and built from scratch in 20 months for about $50 million. This was an unheard-of feat in an era that had seen dirksens (billions of dollars) spent over at least a decade to produce mountains of paper reports and over-head transparencies. If nothing else, the DC-X proved that the aerospace industry (or at least McDonnell Douglas Corporation) could indeed perform in direct contradiction to Cheops' Law.

Cheop's Law is a humorous engineering homily supposedly handed down from the days of the Egyptian pharaoh Khufu and written in hieroglyphics by his scribe: "Nothing *ever* gets built on time or within the budget."

After speeches by an MDC vice president, Col. Pete Worden, and Congressman Dana Rohrabacher of the 42nd California Congressional District, home of the MDC Delta Clipper program at MDC, former astronaut Charles "Pete" Conrad (himself an MDC Vice President) stepped up to the podium. After a very brief series of comments, Conrad waved his hand toward the closed hangar doors and announced, "Well, you've heard all the great things. And the team that put her together is sitting over there. So we'd like you to see the shape of the future. Roll her out!"

The hangar doors slowly slid sideways, revealing in the dark interior of the building a glistening white shape illuminated by floodlights. The DC-X, resting on its yellow "U-Haul" transporter (so-called because it is U-shaped and picks up the rocket at four points around its base) was hauled out into the hazy southern California sunlight by a tractor.

People stood and cheered.

It was a beautiful object. Its white composite aeroshell sported the insignia of the Air Force, BMDO, and the national colors. Lettered across its side were the words, "McDonnell Douglas Delta Clipper Team."

The Team was international and included:

- Aerojet, Rancho Cordova, California—reaction control system.
- Allied Signal Aerospace Co., Torrance, California—actuators and propulsion subsystems.
- Chicago Bridge & Iron Services, Oak Brook, Illinois—tanks.
- Deutsche Aerospace, Munich, Germany—landing gear.
- Douglas Aircraft Co., Long Beach, California—supportability and maintainability.
- Harris Corp., Melbourne, Florida—flight operations control center design and electronic components.
- Honeywell Space Systems Group, Clearwater, Florida—avionics.
- Martin Marietta Astronautics Group, Denver, Colorado—ground support systems.
- Pratt & Whitney, West Palm Beach, Florida—main engines.
- Scaled Composites, Mojave, California—aeroshell.

Everyone present had the opportunity to inspect the DC-X. However, MDC wouldn't let anyone into the Flight Operations Control Center, the "blockhouse" that was the trailer from an 18-

FIGURE 11-1: *The roll-out of the Delta Clipper DC-X at Huntington Beach, California, on April 3, 1993. (Photo by G. Harry Stine.)*

wheeler truck filled with computers and electronics. The MDC ground crew was there in red coveralls, all 20 of them, again a world of difference from the standing army of 10,000 or more needed to fly a space shuttle.

Dr. Jerry Pournelle, Larry Niven, astronaut Buzz Aldrin, and I had ample opportunity to look over the results of our five years of effort on the Council.

Okay, the DC-X wasn't an SSTO. It was a "single-stage-to-twenty-thousand-feet" rocket. But it had progressed beyond paper studies, computer graphics, and overhead transparencies. Now all it had to do was fly.

But not before more tests were performed with it. For example, the DC-X was hung by its nose from an overhead hangar crane and pulled back and forth by ropes attached to its base. This simple action tested the guidance and control system that swiveled the rocket engines. The Team discovered that they had hooked up the system backward, and this was quickly corrected. Other tests showed that the DC-X was ready for hot firings.

Two weeks after the roll-out, the DC-X was loaded on its side aboard a flatbed truck and hauled to the NASA White Sands Test Facility in New Mexico.

This had been built on the west side of the San Andres Mountains by one of my old supervisors at White Sands, John B. Day. Its original task was to test the Apollo Lunar Module rocket engines in the 1960s. It was the newest of the White Sands test stands and was equipped to handle liquid hydrogen and liquid oxygen.

The DC-X propulsion system was hot-fired there during May and June of 1993. The Pratt & Whitney RL-10 rocket engines had been "loaned" to the program by the company to keep the cost down. Ultra reliable and a well-proven design, the engines still had to be tested in the DC-X airframe to make sure the plumbing, valves, and other control devices worked right. Even production airliners undergo ground static testing of their jet engines before their first flights.

The static tests went well except for one glitch. A cloud of gaseous hydrogen collected around the DC-X as a result of liquid hydrogen flowing through the rocket engines to pre-cool them before ignition. When the engines ignited, the gaseous hydrogen cloud also ignited. In videotapes, it's possible to see the spherical flash of this ignition around the ship. In an early test, this burned off the decals on the aeroshell. So the DC-X thereafter had no markings on it, only patches of slightly whiter white where the original decals had been.

These showed up in the photographs of the DC-X that appeared in the first periodicals to grab the DC-X story. Surprisingly, these were not the popular scientific magazines or even the aerospace weeklies.The first stories about the DC-X appeared in two business and financial periodicals, *Business Week* and *Barron's*, on June 21, 1993.

This should indicate that commercial and financial interest runs high when it comes to the potential of the new spaceships.

The implications of this coverage were ignored in the national capital.

The DC-X was getting close to flight.

Would it work?

Or wouldn't it?

TWELVE

"It Works!"

WHITE SANDS MISSILE RANGE in New Mexico is probably the best place in the world to test rockets. It was established by the U.S. Army as a place to fly the 100 German V-2 rockets captured in Germany in August 1945 and brought to the United States. It's a stretch of Chihuahuan desert at an average altitude of 4,000 feet above sea level, an absolutely flat area about 40 miles wide and 100 miles long between the San Andres Mountains on the west and the Sacramento Mountains on the east. It's called the Tularosa Basin and the test range is called White Sands Missile Range because of the expanse of gleaming white gypsum sand that covers a large part of it. The range also includes White Sands National Monument.

White Sands not only saw German V-2 rocket flights from 1946 to 1952 but also the landing of the space shuttle Orbiter Columbia on March 30, 1982.

Major Jess Sponable, the DC-X project officer, decided to locate the flight test site on the south end of the NASA space shuttle landing area known as Space Harbor. A thick concrete pad had been poured to support the cranes necessary to lift the Orbiter onto the back of the Boeing 747 carrier aircraft. Nearby was another concrete pad. Sponable and Paul Klevatt, the McDonnell Douglas DC-X Project Manager, followed the general philosophy of the program: Use available equipment. So the large crane pad became the DC-X launch pad and the smaller one 350 feet away became the landing pad.

Thus was born Clipper Site. It's still called that today at White Sands.

Three miles to the southwest, the Flight Operations and Control Center trailer was located alongside several other trailers housing telemetry receivers and other electronics, plus a mobile home outfitted as an office. The latrine is located all by itself in a building that requires the user to climb a flight of stairs to get to the facilities on the second floor. Not to worry; in the dry desert climate, most water loss occurs unnoticed in perspiration, leaving little to be voided.

Clipper Site is an hour's drive north of the White Sands main base along a road that has mercifully been paved since I first bounced along it in 1951. However, even today, the trip in an Army bus with no airconditioning in desert heat is an adventure. Now as then, rocket flight testing isn't a comfortable profession.

Everything at Clipper Site is temporary. Liquid hydrogen and liquid oxygen are stored in trailers around the launch pad. The "hangar" for the DC-X is a vertical structure made from steel scaffolding and covered by canvas; huge blocks of concrete hold it down in winds, and multiple sets of old rubber-tired wheels allow it to be moved over the DC-X to protect it from the blazing New Mexico sun.

The place has the look and feel of a construction site. It is a construction site; it's where the spaceships of tomorrow are being constructed in the minds of the builders. This is quick-and-dirty rocketry done on a shoestring. If something doesn't work right— and many things don't in an experimental program—it can be quickly changed because a lot of money hasn't been sunk into the sort of massive ground support facilities that exist at Cape Canaveral. Clipper Site is equivalent of the cow pastures the early airplane builders used as airfields with gas trucks instead of fuel farms, pickup trucks instead of specialized towing tractors, and ladders instead of jetways. Clipper Site is an expression of the KISS principle ("Keep It Simple, Stupid!"). When you've got only $59 million to do a job that otherwise would cost $600 million or more in the old, established aerospace way, Clipper Site is the obvious and inevitable result.

It's also a harbinger of things to come. The first spaceports around the world will consist of simple concrete pads and minimal ground support equipment.

Often overlooked is the fact that the flight safety people at White Sands allowed the DC-X to be flown on the range *without a flight safety destruct system aboard.* I know how important this permission was because I was a flight safety engineer at White Sands from 1955 to 1957. It emphasized the fact that the DC-X is indeed the first of a new breed of rockets leading toward true space transportation. Experimental airplanes (or airliners) do not carry in-flight destruct systems. However, since May 29, 1947, when a German V-2 went out of control and landed on Ciudad Juarez, Mexico, 45 miles south of White Sands, no rocket vehicle capable of going beyond the range boundaries had been permitted to fly at White Sands without a device aboard that allows a flight safety officer to destroy the rocket and prevent it from landing elsewhere.

The DC-X sits ready for flight on its own launch stand. Four pillars support the ship at its base. Running up inside these pillars are the rocket propellant feed lines, the ground electrical power lines, and the other service feeds that provide pressurized nitrogen and other victuals for the DC-X. No mobile skyscrapers are used to permit the ground crew to reach the upper hatches and service panels; extension ladders and industrial hydraulic lifts were rented and pressed into service instead. The original "mobile skyscraper," the shipyard gantry crane that was modified to allow easier work on German V-2 rockets, still sits in the original White Sands Army Launch Area, now a historic landmark. Little did we realize that the old gantry would breed the huge structures now at the Cape.

Early in August 1993, the DC-X and Clipper Site were both ready for the flight test series. On the White Sands test stand, the DC-X had undergone nine hot firings of its rocket engines, and all flight conditions possible were tested first on the ground during these firings.

Most important from the standpoint of the experimental test objectives of the DC-X, two hot firings were made within eight hours. The three-person flight crew and the 25 individuals on the ground crew—including program managers, test support people, and maintenance personnel—learned how to preflight the DC-X in two to three hours.

Only the Delta Clipper Team members witnessed the first flight of the DC-X on Wednesday, August 18, 1993. It was two years and two days from the date that MDC had received the DC-X contract. Hundreds of others saw it via closed-circuit satellite television link to the MDC Huntington Beach plant.

Late in the afternoon, the ground crew retired to the flight operations control area. Liquid oxygen and liquid hydrogen were loaded by remote control into the DC-X tanks. The DC-X on-board computer ran through its pre-flight tests and reported to Pete Conrad in the control trailer that everything looked "notional." Conrad received range clearance, comparable to getting takeoff clearance from the control tower for an airplane. From his computer terminal, Conrad told the DC-X to fly.

FIGURE 12-1: *The people who did it at the press briefing before the first public DC-X flight on September 11, 1993. Left to right: Charles "Pete" Conrad, Paul Klevatt, Major Jess M. Sponable, Col. Simon P. "Pete" Worden, and White Sands Commanding General Richard W. Wharton. (Photo by G. Harry Stine.)*

At 4:43:53 P.M. MDT, the engines ignited at 30% thrust.

The DC-X computer checked the rocket. All engines were running. Everything was in the green. It commanded throttle-up.

Because of the partial fuel load, the DC-X rose quickly and majestically off the launch stand.

At an altitude of 150 feet, it stopped climbing and hovered in mid-air!

We're used to seeing a rocket lift off and disappear into the sky overhead. People watching the videotape of this first DC-X flight are astounded when the vehicle comes to a halt in mid-air.

The rocket engines swiveled a few degrees on command from the autopilot and, 13 seconds after takeoff, the DC-X began to move sideways at a brisk walking pace, holding its altitude of 150 feet.

After moving about 350 feet to the southwest, the DC-X stopped again in the air as its engines swiveled to stop its sideways movement. The on-board Global Positioning System (GPS) was receiving signals from the GPS satellite system in space and determined that the DC-X was directly over its intended landing point. The computer told the DC-X to land. The engines throttled back, and the descent began. At 100 feet, the retracted landing legs extended from the four corners of the base.

At a height of 30 feet above the concrete landing pad, the four jets began kicking up dust. The landing radar—similar to that used on the Apollo Lunar Module that Pete Conrad landed on the Moon 24 years before—piloted the DC-X carefully toward the ground where it landed in a cloud of dust colored by the exhaust jets.

When the four landing legs touched the concrete pad and compressed under the 20,000-pound weight of the vehicle, the computer shut down the engines and began to go through its post-flight procedure. As the DC-X became visible again, the call came from Pete Conrad, "Touchdown! Touchdown! Weight on gear! Weight on gear! Engine shutdown!"

And someone in the flight control center trailer shouted, "All ri-i-i-ight!"

It really wasn't much of a flight in terms of what rocket vehicles usually do in terms of speed and distance flown. It wasn't supposed to be. A McDonnell Douglas spokesman claimed it was "equivalent to the high-speed taxi test of a new airplane." But in such a preliminary taxi test, daylight doesn't appear between the airplane's wheels and the runway. The DC-X had risen to three times its own length and landed.

That little "bunny hop," as Pete Conrad described it, had far greater impact than any rocket flight in the last 25 years.

There were two reactions to it in Washington:

1. "All ri-i-i-ight! *It worked!*"
2. "Ohmygod! *It worked!*"

The videotape of the flight even made the *NBC Nightly News* where Tom Brokaw seemed as amazed as nearly everyone else.

Except for a scorch of one side of the composite nose cone caused by the burn-off of the vented hydrogen at engine ignition (the nose cap was replaced), the DC-X could have been readied for its second flight within two days. But the flight crew wasn't ready. They were dead on their feet. They hadn't had a day off, even on weekends, for a month. And they'd worked straight through for 48 hours before the "bunny hop" taking care of all the hectic details. Sponable and Klevatt wanted to look at the videotape and telemetry data. And they wanted to scrub down the aeroshell with green industrial cleaner and a stiff brush to get all the scorch marks off.

The quick turnaround—flying, then flying again in two days—was one of the later tests scheduled in the series.

However, if the Delta Clipper Team wasn't ready for a quick turnaround, the opponents of the program were indeed ready to do just that.

Some high-level NASA people did what pilots call an "immediate one-eighty."

THIRTEEN

The Huntsville Meeting

THE SPEED WITH WHICH the Gun Club reacted to the successful DC-X flight test on August 19, 1993 was astounding. Their responses had to have been thought out well ahead of time; *no organization* as large and powerful as the Gun Club can react as quickly as it did without preplanning.

Their alternate position—"We told you it wouldn't work!"—was immediately abandoned.

On Thursday, August 25, 1993, *one week* after the DC-X flight, NASA at the direction of Administrator Daniel S. Goldin told the aerospace industry to send two representatives each to a meeting at NASA's George C. Marshall Space Flight Center in Huntsville, Alabama, on the following Tuesday, August 31, 1993. Invitees included such companies as Rockwell International (builders of the space shuttle), and Martin Marietta Corporation (builders of the Titan expendable launch vehicle family). Also invited were people from Boeing and Lockheed. Any others who wanted to attend could do so. Many did.

At the same time, Goldin privately told Col. Pete Worden that he was *very* unimpressed by the DC-X flight because "it contributed nothing to the technology base." That isn't what he was saying a few months later.

At this point, it was learned that there had been an ongoing internal NASA study called "Access to Space." The various NASA centers had been given the job of looking at various new launch vehicle options. The person heading the Huntsville study group was Gene Austin. His team chose to study what they termed "single-stage, fully-reusable launch vehicles." The numbers came out looking good.

NASA's Langley Space Flight Center, the original home of manned space flight before NASA Johnson Space Center in Houston took over, also discovered that an SSTO was "now feasible."

In issuing the invitation to the Huntsville meeting, Goldin set forth the following plan:

He said NASA intended to build an autonomous, robotic, unmanned launch vehicle that would be single-staged and totally reusable. It's mission was to carry cargo only and resupply the space station. This rocket, later in slightly modified form tagged X-2000, would not be built and flown until the year 2000. In the meantime, NASA would spend $100 million to $300 million per year doing technology studies to ensure that "the technologies were well-understood and in hand" before construction of the first X-2000 began.

It looked like NASA was going to build a "perfect rocket" like the space shuttle.

According to Goldin, NASA was "leaning strongly" toward a Vertical-TakeOff-Horizontal-Landing (VTOHL) configuration like the space shuttle with which NASA had experience. Lockheed and Rockwell International both were on record as favoring that approach. Vertical-TakeOff-Vertical-Landing (VTOVL) concepts like the Delta Clipper would be welcome, but VTOHL was favored.

NASA would do some cost-sharing with contractors, and a consortium effort was expected. Goldin said the formal Request for Proposal would be issued on October 1, 1993, with industry responses due on November 1, 1993, and contracts issued on December 1, 1993.

This is terribly quick action for NASA as well as for the aerospace companies. A formal response to a Request For Proposal submitted within 30 days usually means that the response has already been written and cleared through Contracts, Legal, Engineering, etc., and the photocopy machines are ready to run. Furthermore, a smart NASA contractor doesn't deviate very much from the baseline set forth by the customer, especially if said contractor has spent some time convincing the customer that the contractor has the One Good And True Way to do the job.

If this NASA X-2000 plan sounded like Son Of Space Shuttle or Shuttle II and Never-Launch-System (NLS) combined with the canceled X-30 National AeroSpace Plane (NASP), it gets worse in this regard as the story unfolds.

Where was NASA going to get the money for X-2000?

If the space station died, $100 million could easily be shifted around. NASA has a lot of pockets in which to squirrel away odd pieces of small change like $100 million.

However, this may be a moot point. NASA was *the* agency that's supposed to develop space vehicles. If NASA announced that it intended to develop SSTOs, the two appropriations committees of Congress would immediately ask themselves why they should continue to fund SSTO development in the Department of Defense through the Advanced Research Projects Agency (ARPA). ARPA had sent signals that it didn't want SSTO. After all, ARPA had just screwed up two space launch vehicle programs—Pegasus and Taurus—and didn't want another space launch vehicle program. That situation pretty much killed the possibility of doing SSTO as an experimental or X-vehicle program within the Department of Defense. If the Air Force did it, SSTO would have to be projected as an operational vehicle.

But without a "mission requirement," there is no justification for requesting money for an operational vehicle. The Air Force generals weren't excited about SSTO because, although it could carry a pilot and co-pilot, it didn't have wings like an F-15. Most of the high brass were fighter pilots.

It appeared that this was a masterful move on the part of NASA to take over all SSTO development once DC-X and NASA internal studies clearly indicated that SSTO would work after all. From the Congressional viewpoint *as well as* that of the Department of Defense, this would relieve a lot of pressure on the defense budget in a time of post Cold War meltdown of the armed forces.

The timing was perfect because the budget battles for the next fiscal year were already under way.

It was a brilliant bureaucratic move: Kill SSTO by adopting it as NASA's own and turn it into a technology study program. No more bending tin and flying something that could fail, thus threatening the budget.

This plan came from an agency that had just lost Mars Observer, was having trouble keeping the space station funded, and had allowed the cost of flying a space shuttle mission to exceed a dirksen.

Furthermore, turning a project into a technology development program had killed every projected launch vehicle program in the last 25 years.

But not before NASA could spend a couple of dirksens on "SSTO technology development studies." NASA had its own house to worry about if it lost the space station and if its budget were cut. NASA managers believed they had to protect their people and their contractors as well as the rice bowls from which they all fed. NASA desperately needed an ace-in-the-hole to save its high-tech jobs programs if one of its projects went toes-up.

And that is what the national space program had become: a high-tech jobs program.

Col. Pete Worden went down to Huntsville on Monday, August 30, 1993, to sit in on the NASA meeting. The Air Force officer believed that the best he could do was to work with NASA, given the environment in the Department of Defense.

Worden was therefore surprised when General Carns, Air Force Vice Chief of Staff, tracked him down in Huntsville on the evening of August 30, 1993 and pointedly asked on the telephone, "Colonel, what the ____'s going on?" When a four-star general says that to a bird colonel, it gets the colonel's attention. Carns made it clear that the Air Force *and* the administration wouldn't support what NASA was doing because the space agency had jumped the gun and could potentially kill all hope of doing anything. Worden was told that the Air Force Chief of Staff, General McPeak, had planned to support some kind of follow-on SSTO effort as a low-profile cooperative arrangement with NASA, but couldn't openly say so.

What set off the fireworks in Air Force Headquarters was a combination of Goldin's fire-fighting management style, the successful first flight of the DC-X, and the way that NASA Marshall Space Flight Center wanted to build the X-2000, if it ever did, as a totally in-house effort. The Center would act as its own prime contractor, designing the X-2000 down to the last rivet and pipe joint. From the aerospace contractors, Marshall would buy components built to these design requirements and tight NASA specifications. The time and cost estimates boasted three years to first suborbital flight of an X-2000 prototype (not an experimental vehicle) at an estimated cost of $350 million. The Air Force was not included in this plan. Maybe the Air Force couldn't do the SSTO program themselves, but they didn't want to be left out if it was done by some other agency.

The Huntsville meeting started with Gene Austin, the NASA Access to Space study leader, talking for an hour and a half about the study's conclusions. Then the briefers presented NASA's technology roadmap—SSTO, all rocket, vertical takeoff, horizontal landing, and using a new and untried rocket propulsion scheme called "tri-propellant."

As discussed earlier, one of the problems of building an SSTO involves the performance of its rocket engines—specifically, the need for altitude compensation. The early proposals in the SDIO Phase I study envisioned using plug nozzle rocket engines. But these weren't off-the-shelf. The MDC engineering make-do for the DC-Y Delta Clipper would have used extendable expansion skirts on half of the rocket engine bells. Yet another solution, known to most of us but dismissed as not being off-the-shelf, is the rocket engine that uses three propellants—liquid oxygen (LOX) and either liquid hydrogen (LH2) or something akin to jet fuel (RP-1). During the flight to orbit, the tri-propellant engine starts out burning liquid oxygen and jet fuel, then shifts to liquid oxygen and liquid hydrogen at altitude.

A tri-propellant rocket engine has never been developed in the United States. The Russians were on the verge of developing one, the RD-710, when the Soviet Union collapsed.

NASA said the proposed X-2000 rocket would use the Russian RD-710 engine although it hadn't yet been tested. However, this involved a foreign source that might place the entire program at the mercy of foreign policy or a change in Russian attitude. This didn't seem to be a prudent move to the Council members who heard about it later that day.

The NASA technology map proposed a multi-billion-dollar program to make sure all the critical technologies were ready by the year 2000 when the X-2000 would start flying back and forth to orbit.

Members of the Council had heard this before. It was space shuttle all over again. It confirmed our fears that NASA would kill the SSTO by turning it into the ordinary ten-year ten-dirksen paper studies program.

The whole NASA X-2000 program had been thrown together in two weeks. It seemed to be a blatant attempt by NASA to corral the follow-on to the DC-X before it started breaking rice bowls.

Across the Potomac from the Pentagon on Capitol Hill, the NASA move found its supporters and detractors. A typical pro-NASA supporter was Dr. Harry S. "Terry" Dawson, engineering advisor on the staff of the House Science, Space, and Technology Committee. Dawson didn't like SSTO, was satisfied with the status quo, and became a very active opponent on the Hill. He was in a crucial position on the Committee. Some of us wondered why he was so anti-SSTO. It turns out he wasn't. He was basically opposed to building and flying experimental vehicles.

Who was this man? Why was he behaving in this manner?

It turns out that Dawson had been an amateur rocketeer as a young man. He'd built his own rocket engines. Dawson admitted in a luncheon speech that he'd built and tested 200 rocket engines, trying to develop one that would allow him to break the so-called altitude record for amateur rockets. This marked him immediately as a person who wants to collect one more data point to build the perfect device, the "scientific type" who will fiddle and tweak and "improve" on something endlessly without a commitment to go ahead with the less-than-perfect device.

Dawson had a Ph.D. in engineering from the University of Maryland and had spent his career doing what at best might be described as "paper engineering." He spent 16 years as an engineering and management consultant for various think tanks in the Washington area. His resumé shows that he'd worked for long list of Washington bureaus and offices including NASA, FAA, Urban Mass Transit Administration, the Environmental Protection Agency, the Arms Control and Disarmament Agency, the National Security Council, and the Office of Technology Assessment. He also did work for the Defense Advanced Research Projects Agency (DARPA, now ARPA). His resumé also states that he worked with "many elements of the intelligence community." On the side, he was past president of the National Space Club, served on the board of directors of the National Space Society, and was a member of the "policy making body" of the American Institute of Aeronautics and Astronautics. Obviously, Dawson was very well-connected in Washington.

Dawson had supported the development of the air-launched Pegasus and the expendable Taurus space launch vehicles. Both of these were less than successful, and Dawson found himself burned badly as a result of his advocacy. This threatened his position with the House Committee on Science, Space, and Technology. So he pulled in his horns.

Therefore, he was unwilling to support the SSTO program. He had friends on the Hill and elsewhere who agreed with supporters in NASA on insisting that all technology be in place before actually applying it.

This can't be done.

The very word "technology" derives from the Greek word *teckne* which means "know-how." The only way to obtain know-how is to do something until you make it work. Good engineering has always proceeded from failures in design or meeting requirements.

On the other side were SSTO supporters like Tim Kyger who was at that time on the staff of Congressman Dana Rohrabacher (R, CA). Tim Kyger is a space advocate in his guts. He was a tiger (and I've often wondered, given his surname, why he didn't pick up that obvious nickname) in the California activities of the L5 Society before it merged into and was swallowed by the National Space Society.

Kyger felt there would be a reluctance in Congress to give everything in the launch technology pile to NASA. The space agency had lost Mars Observer and the NOAA-13 weather satellite. However, he also pointed out that the whirling knives of politics were spinning again. Keeping SSTO in the Department of Defense might mean the Air Force giving up one of its 20 active fighter wings to fund a launch vehicle development that the generals didn't really want.

This background is important in light of what happened next.

Air Force Headquarters worked its wiles on the Office of the Secretary of Defense who then passed the word along to the Executive Office of the President. Sources are vague at this point concerning who said what to whom. We do know that the office of the Air Force Chief of Staff had its oar in the water. However, results were what counted at that moment.

About a week after the Huntsville meeting, NASA Administrator Goldin reportedly got a telephone call from the White House saying, in effect, "Not yet. Be patient. The time will come. So cease and desist for now."

The threat of NASA taking over and killing the SSTO had temporarily been blunted, but another threat immediately loomed on Capitol Hill during the appropriations fracas for the next fiscal year.

FOURTEEN

The Opposition
Heats Up

THE STAFFERS ON CAPITOL HILL, some of them anti-SSTO types, weren't at the infamous NASA Huntsville meeting. Between August 23 and September 3, 1993, Dr. Terry Dawson of the House Science, Space, and Technology Committee, Gary Sojka of the Senate Intelligence Committee, Bruce MacDonald of the House Armed Services Committee, Jack Mansfield of the Senate Armed Services Committee, Larry Cox of the House Intelligence Committee, and the Air Force Legislative Liaison, Steve Jacques, were out touring the Air Force Space Command and various aerospace companies on a "Space Launch Oversight Trip."

The trip report consisted of a series of overhead transparencies about the present-day state of U.S. space launch capabilities and philosophies. The sorry condition of space transportation hadn't been ignored in Washington, but few people there seemed to know what to do about it. The report confirmed this.

Several "bottom-up" reviews had already been made in 1993, and all of them generated three basic options:

(1) maintain the current fleet of expendables and the space shuttle, making incremental improvements over the years;
(2) develop a new expendable launch system aka ALS/NLS/"Spacelifter"; or
(3) develop a new launch system such as SSTO using what

is termed "leap-frog" technology (a denial of the fact that SSTO technology is by and large already available).

The trip report reported a total lack of consensus in Washington regarding which of these three options should be chosen. The Department of Defense wanted #1 while the Air Force chose #2 and NASA opted for #3 without specifying SSTO as a solution. Given this lack of agreement, the Clinton White House tackled the problem by initiating yet *another* "bottom-up" review that may never have been completed because no one on the Council has ever seen it.

Another finding in the trip report was something many of us already knew: The United States was building the space-going equivalent of race cars—performance-driven, fragile, expensive, complicated, difficult and time consuming to manufacture, and with steep performance decline when out of tune (used outside of narrow limits). On the other hand, the Europeans and Russians were building space trucks that are cost- and reliability-driven, durable, cheap, simple, rugged, easy to manufacture, and with wide performance tolerances. The Eurospace Ariane IV requires a launch crew of only 100 people and can be put on the pad and flown within ten days while the American Titan IV demands more than 1,000 people and 100 days on the pad. The Russian boosters were even better because they're built in one-third the time with one-third the people, spend only a few hours on the launch pad, and are so rugged they can be launched during a blizzard.

The consequence of this situation is clear considering the fact that the United States' share of the commercial space launch market declined from 100% in 1972 to only 30% in 1992.

The travelling staffers decided a new philosophy was required. Space launch should be viewed as a *service* to be managed like airlift, sealift, a truck line, etc. This was seen as a refreshing new approach.

The report recommended that "the best way to implement this "major philosophical change" was to *appoint a space launch czar!* Such a space access dictator would have total budgetary and policy control over all launch vehicles as well as the ability to dictate payload requirements to users, impose major personnel reductions throughout the space launch community, restructure the operation of all national launch ranges, and form consortia in which aerospace companies would share information and costs.

This was the sort of highly centralized bureaucratic control that worked during the Cold War to develop such weapons as ballistic missiles and nuclear submarines. But it amounted to a continuation of a government space program with restricted access. The U.S. Postal Service is a close analog to what such a space organization would be like, but with no opportunity for FedEx or UPS to offer competition.

Fortunately, this report had little impact and was dismissed by knowledgeable space transportation individuals and organizations. However, it remains an indication of the mind-set of important non-elected, appointed, and non-responsible people who hold critical reins of power on Capitol Hill and in the executive branch of government. It should be taken as Lesson #1 in the true source of power in the federal government: the congressional staffers and the civil service bureaucrats, none of whom are elected.

In the meantime, the DC-X flew *again* on September 11, 1993, as reported in Chapter One. And again on September 30, 1993, a flight that revealed the stout design philosophy of the DC-X.

On the third DC-X flight, the RL-10 engines ignited, but a helium bubble in the feed line for one engine caused it to remain at low thrust while the others came to takeoff thrust on command of the on-board autopilot/computer. The DC-X lifted off and began to tip to one side because of the unbalanced thrust. The control system detected this, swiveled the other engines, and increased their thrust to compensate. The rocket slewed sideways, bathing the launch stand in the jet exhausts of the working engines. Then the recalcitrant engine came to full thrust. The DC-X righted itself and flew through the rest of its test. Other than a scorched launch stand that was subsequently cleaned up, the DC-X itself was in perfect condition. The same could not be said of the test crew, whose members gained a few more grey hairs that day.

Back in Washington, the SSTO opponents knew that the one sure and positive way to stop a government program is to cut off its funding. That was the nature of their next attempt to kill the SSTO program.

A brief review of the government funding process will clarify the process behind several of these attempts that have occurred since September 1993.

The congressional funding process has two phases. The first is "authorization" followed by "appropriation." Authorization is like drawing up a shopping list for the coming year. Appropriation can be considered going through the list to determine how much of each item, if any, to buy, given the amount of money available. Authorized budget items are often reduced or deleted in the appropriations phase but seldom increased. No new items are added to the list in the appropriations phase.

Throughout this process, members of Congress and their personal staffs are subjected to pressures to do or not do something, increase or decrease a budget item, or add/delete line items by a legion of lobbyists, some registered as such and some not. Not all lobbying takes place in the halls and offices of the Capitol and the congressional office buildings. A great deal of it goes on at the personal level *between staffers with personal agendas*. This is especially true for committee staff members. Most of an authorization or appropriation bill is written not by the congressperson (who hasn't got time to do it) but by their staffers who have been told to do it or who have taken on the job. Or the bill can be written by the committee or subcommittee staffers. The process is so complex that a single congressperson can't know or follow all the details in all the various bills unless that elected official happens to have an interest in a specific budget item (most do when it's related to patronage or pork). Congresscritters rely on staffers to provide the answers to such questions as: "What does this mean? Should I vote for or against this?"

On September 14, 1993, when the National Defense Authorization Bill came before the Senate for floor vote, senators Peter Domenici and Jeff Bingaman from New Mexico submitted an amendment to fund the continuation within the Department of Defense of the DC-X flight tests and the design and building of the follow-on SSTO test vehicle. The House had already authorized $75 million. Similar funding seemed assured in the Senate because it was a bipartisan proposal of modest scope by congressional standards.

It was a modest request because the Department of Defense has a long history of funding experimental vehicles such as the X-15 and other technology risk-reduction projects. It was also modest because the Department of Defense at that time provided far larger chunks of money out of its budget for such nondefense programs as breast cancer research, the National Defense Center for Environmental Excellence, prostate disease research, environmental impact on Indian lands, urban youth programs, US-Japan management training, the Coal Utilization Center, assistance to local educational agencies, and the National Center for Advanced Gear Manufacturing, for example. According to a study released on March 30, 1994, by the Heritage Foundation from which this list was excerpted, 4.6 dirksens were appropriated in the 1993 defense budget for nondefense purposes. This is enough to maintain two Army divisions, buy a nuclear-powered aircraft carrier, or build a defense for the US and its allies against ballistic missiles that could be launched by North Korea and Iran. Thus, a $50 million request for SSTO development is almost lost in the noise.

The Senate rejected the SSTO amendment by a 66–33 vote.

What happened?

Just before the vote on the Domenici–Bingaman amendment, a "fact sheet" was distributed in the cloak room to "guide" the Senate deliberations. This had been prepared by members of committee and congressional staffs as well as some officials in the Department of Defense, perhaps including those who participated in the Space Launch Oversight Trip in August and September. Although I have a copy of this "fact sheet," I don't know who wrote it or handed it out because it's unattributed.

The text of this "fact sheet" reads as follows:

DOMENICI SINGLE STAGE TO ORBIT

Arguments for:
    —*Has captured public imagination*
    —*Efficient new way of doing business*
    —*Could possibly make space flight as common as air travel*
    —*Aims to reduce cost from $5000 per pound to $50*

Arguments against:
    —*Program will cost $6 billion*
    —*700 ton spaceship flies up and back*
    —*Risks are very great. Requires new engines, new engineering*
    —*If weight growth is even 1.5%, payload would be zero*
    —*Wrong vehicle for DoD*
    —*Optimum for manned flight, millions of pounds per year*
    —*But DoD needs less than 100,000 pounds per year*
    —*Possibly right vehicle for NASA*

The Space Transportation Association, the Space Frontier Foundation, the National Space Society, and several other organizations immediately issued *attributed* white papers refuting the outright lies in this "fact sheet." Here are a few of the rebuttals:

- The $50 per pound cost was never claimed by anyone in the program.
- The SSTO program and its follow-on will cost only a maximum of $400 million per year over a three-and-a-half-year time period. This is far less than $6 billion. The follow-on vehicle will also answer all open questions and pave the way to significant cost reductions.
- The 700-ton weight (1.4 million pounds) is more than a factor of 5 larger than the follow-on experimental vehicle and larger than that required for the DC-1 Delta Clipper operational commercial spaceship.
- Risks exist in any project, and that's why a series of experimental vehicles is planned. Independent studies by other aerospace companies and consultants have concluded that SSTO can be done with existing technology and

materials. If it turns out that it can't, the program can be cleanly ended at once.

- The weight margin of 1.5% is off by an order of magnitude. The Delta Clipper DC-Y design has a margin of 15% and will still lift an Atlas-sized payload to orbit if the weight grows by 20%. The 1.5% margin statement is not only incorrect but dishonest.

But it was too late. The killer was the false quote of six dirksens as the price tag. The amendment failed.

This sort of skullduggery, perhaps illegal, is certainly unethical.

However, $50 million was put back into the conference bill, thanks to fast work on the part of supporting staffers and space advocacy groups.

Over in the House of Representatives, other high jinks were going on. Because staff members are the ones who write the language of the bills and even do the compromising with other staffers, anti-SSTO language kept being inserted into the authorization and appropriations bills and their conference bills as well.

A clever one was language ordering the Advanced Research Projects Agency to run the SSTO program themselves, which would allow them to disband the DC-X team because ARPA has no great interest in actually flying rockets. Then ARPA could piddle away the money and end up with perhaps a new supercomputer having "DC-X" written on the side.

On the other end of the subtlety scale was a recommendation to turn the SSTO program into a "thoughtful, long-term research and development program" specifically aimed at a space shuttle successor or the rejuvenation of the "Never Launch System." This would have made SSTO into another X-30 National AeroSpace Plane (NASP) and would never produce a single actual flight vehicle.

Either approach would create a "National Aerospace Jobs Program" or a "porklifter" project. These are as likely to produce an economical, reliable, reusable commercial spaceship as giving a set of crayons to a child and getting a Rembrandt as a result.

Throughout the remainder of 1993, a stream of negative comments continued to pour from the SSTO opponents. For example, in the September 20–26 issue of *Space News* on page 17, an article entitled "DC-X: Publicity Stunt or Giant Launch Technology Leap?" appeared under the byline of Leonard David. In it, Dawson was quoted anonymously (but he was later discovered because this man's cover was gone by that time): "The corporate marketing guys who thought up the Delta Clipper ought to win a national prize. If you make a list of all the things you've got to do to get to orbit in a single stage, the DC-X has almost none of them. . . . You can fly higher than this thing in a Cessna. . . . It's beautifully contrived and it's a stunt. They're marketing a lot of snake oil." Dawson went on to state that the limited technology developed by the recently tested DC-X did nothing to bring SSTO closer to reality but certainly did bring McDonnell Douglas considerable attention from the press.

The McDonnell Douglas Delta Clipper Team really didn't want attention from the press. They wanted to build the DC-3 of space. Furthermore, Gaubatz and his team were having to do it without the full-blown *public* support from the MDC officers or Board of Directors. MDC wasn't about to screw up their relationship with NASA on the space station or be perceived as putting a lot of corporate support into a project that might send jittery stockholders running to their brokers. Press attention? What the Delta Clipper Team wanted was to do what they said they could do.

Fortunately, they didn't get media attention because what happened next would have looked really bad.

The government fiscal year ends on the final day of September each year. Sponable had just enough money left in the sock to support a couple more flight tests before the next fiscal year's funds became available once the appropriations bill was passed and signed, hopefully a few weeks hence. The bill contained $50 million for the SSTO program—$5 million to complete the DC-X flight tests and $45 million for the design of the follow-on SSTO. But the money wasn't there yet.

Late in the afternoon of September 30, 1993, one of the White Sands Missile Range accountants presented the DC-X office at Clipper Site with a large bill for accumulated range support services. The White Sands accounting office wanted to be reimbursed out of current fiscal year funds. They were just keeping the paperwork clean. But this wiped out the remaining flight test funds.

The DC-X was grounded.

Not because of inadequate technology but because the program ran out of money.

FIFTEEN

Sitting in the Hangar
and on the Funds

THE DELTA CLIPPER TEAM at White Sands had no recourse but to wrap the DC-X in plastic sheets, roll the hangar over it to protect it, shut down Clipper Site, and go home to Huntington Beach. Further flight testing of the DC-X would have to wait until money from the next fiscal year budget became available. Even in a dormant condition, the project would incur thousands of dollars in site maintenance and security costs.

But Sponable and Gaubatz refused to treat the hiatus as a loss. They decided to monitor the DC-X in its stored state to learn what happens when a reusable rocket isn't used for a long time. In the desert, airplanes can be stored for months or even years if certain parts such as plastic windshields are protected against the intense sunlight and various openings are sealed against sand and nesting birds. But no one knew how long a reusable rocket would last because the DC-X was the first. Since the "X" in DC-X stands for "experimental," the delay was turned into an experiment of its own.

But Washington politics didn't shut down.

On October 18, 1993, the draft of an internal white paper was circulated around NASA Headquarters. It was written by Dr. Ivan Bekey, formerly of Aerospace Corporation and now on NASA Headquarters staff. Bekey is widely known and highly respected for his careful work. The technical analyses used by Bekey in the white paper were performed by Richard Powell, Roger Lepsh, and

Douglas O. Stanley of the Vehicle Analysis Branch, Space Systems Division, NASA Langley Research Center.

The title of Bekey's draft white paper was, "Why SSTO Rockets Are Now Feasible and Practical."

NASA had now made a 180-degree turn from a public position of nonbelief a year before. It was obvious to those of us whose fax machines spat forth this document—smuggled out of NASA headquarters—that NASA was indeed setting the wheels in motion to take over all SSTO development. Given the Big Project mentality in NASA, this could only mean that SSTO was under serious consideration for the space shuttle replacement, Shuttle II.

The final draft of Dr. Bekey's white paper officially appeared on January 4, 1994, after it had been "staffed" around NASA Headquarters to ensure that it was indeed the new party line.

Among the white paper's conclusions, reported here verbatim, with the typical governmental grammar, syntax, and punctuation:

> This paper has shown that the reasons why many people consider SSTO rockets to be unattainable, or at best very risky and marginal ventures, are based on outdated and in some cases erroneous perceptions. While most of these perceptions were appropriate in the past, when the limitations of then-existing technology rendered them prudent, they are no longer valid given the technologies that can be available in a few years. . . . The net result is that the risk in undertaking the development of SSTO rocket vehicles in the near future should not be substantially greater than that associated with launch vehicles using "conventional" technologies. But the required advanced technologies must be matured in a focussed technology/advanced development program, and demonstrated in an X vehicle, for this result to be obtained. The maturation and demonstration process can and should consist of a highly focussed program lasting at most 3–5 years. . . . Thus the concerns should shift from ensuring that SSTO rocket vehicles are feasible, to ensuring that an adequately funded focussed technology development and

demonstration program is carried out. This will then
enable confident and responsible consideration of decisions to begin vehicle development.

Between the draft and the final paper, sentences were added relating
to the advantages of running a focussed X-vehicle development and
demonstration program taking no more than three to five years.

This is now NASA's official position with regard to SSTO. It
was obvious that NASA wanted SSTO once the DC-X showed that a
reusable rocket was feasible.

If this was the case, it was then a matter of seeing to it that
NASA did not turn it into a ten-dirksen, ten-year project after all.
The way to keep that sort of thing from happening in any government agency is to ensure that Congress maintains close oversight. It
looked like the work of the Council wasn't over.

The Bekey white paper helped silence the SSTO adversaries. It
certainly exposed the lies of the infamous "fact sheet" that caused
the Domenici-Bingaman funding amendment to fail on the floor of
the Senate a month before the draft paper leaked.

The Ballistic Missile Defense Organization's handling of the
SSTO project was further boosted by the success of another of their
projects, Clementine. This was a bare bones spacecraft using technologies developed in the ballistic missile defense program.
Clementine was done fast and cheap. It was operated out of a "mission control" located in an undistinguished old brick warehouse,
the "bat cave," a former National Guard armory in Alexandria, Virginia, by a crew of only 55 people. Clementine digitally imaged
100% of the surface of the Moon under constant geometry and
lighting conditions and in 11 different wavelengths.

One piece of Clementine data was extremely important to future
space activities:

Frozen water was discovered in one of the craters near the lunar
pole.

This means that we will have a source of hydrogen and oxygen
for rocket propellant when we get there.

The success of the Clementine program didn't stop Dawson
from continuing an attempt to gain some power over the SSTO

program in NASA. SSTO apparently threatened his turf. As late as December 3, 1993, *Space Business News* reported that Dawson, during a speech to the Aerospace States Association, continued to push hard for a "space czar" who would have dictatorial control over all space launch vehicle design, development, and operation. Perhaps Dawson was angling for such a position himself. In making this speech, Dawson was billed as part of a group of staffers representing seven of the ten congressional committees having jurisdiction over space launches who at that time were attempting to draft omnibus legislation to deal with this. Dawson derided American launch vehicle and satellite builders for their inefficiency compared to the Russians. "I have an M.B.A. degree, and I don't remember any case studies of the Russians being held up as a model of efficiency." (Dawson does *not* hold an M.B.A., according to his own official biography. His only listed Master's degree is in Engineering Administration from George Washington University.)

Lest I be accused of personally disliking Dawson, let me say that I have never met the man. I only know him from what he has *done* and *the way he has done it*. I have seen one photograph of him taken shortly before I wrote this. He may be a "space cadet" as he claims, but I don't believe he's *my* kind of space cadet.

As for the funding to continue the DC-X flight tests, it was hung up in the budget appropriations process until November 10, 1993, when the Defense Appropriations bill was sent to the White House for signature. A total of $50 million was appropriated to the Advanced Research Projects Agency specifically authorizing ARPA to subcontract the SSTO work to the Air Force but was not forbidden to subcontract it to BMDO. Of this sum, $5.1 million was appropriated to finish the DC-X flight tests and $40 million to begin the design of the SSTO follow-on.

The money was thus made available to ARPA.

And there it sat.

Just because a government agency has been given funds doesn't mean that it has to spend them. And ARPA did not.

Specifically, one person at ARPA was responsible for this.

Dr. Gary L. Denman was appointed Director of the Advanced Research Projects Agency on March 15, 1992, after serving as agency deputy director from 1990 to the date of his promotion. His B.S. is in mechanical engineering from the University of Cincinnati as are his M.S. and Ph.D. degrees from Ohio State University. According to his resumé, he never worked for commercial industry in the private sector but spent his career as a civil service engineer at Wright-Patterson Air Force Base managing programs related to advanced materials and manufacturing technologies. At ARPA, Denman was responsible for "research, development, and demonstration of concepts, devices, and systems that provide highly advanced military capabilities."

Denman wasn't quite so sure he wanted the SSTO program.

ARPA had besmirched its reputation by backing two supposedly cheap, fast, simple space launch vehicle programs: Pegasus and Taurus. Pegasus was air-launched—at first from a B-52 and later from a converted Lockheed L-1011 TriStar wide-bodied jet airliner—and flew part way to orbit. Taurus was a sophisticated version of a big, dumb booster made up of Minuteman III ballistic missile motors. Both ran into trouble technically, financially, and operationally. Denman didn't want any more space launch vehicle programs because he'd been burned by supporting these two launch vehicles.

So Denman sat on the SSTO funds.

When queried about this by members of Congress and their staffers, Denman told them he didn't want to go against the wishes of his boss, Under Secretary of Defense for Acquisition, Dr. John M. Deutch.

From his resumé, Dr. Deutch appears to be a typical "political scientist." With degrees in chemical engineering and physical chemistry from the Massachusetts Institute of Technology, he has a long list of academic positions, including Provost of M.I.T. In Washington, he was one of Defense Secretary McNamara's "Whiz Kids" from 1961 to 1965 and lists service to the Bureau of the Budget, the Urban Institute, the National Science Foundation, the Army Scientific Advisory Panel, the Defense Science Board, the National Security Council, and the Defense Science Board. A member of the Trilateral Commission, he holds memberships in the American Academy of Arts and Sciences and the Boston Museum of Fine Arts. The resumé also lists him as a member of the boards of directors of Perkin-Elmer Corp., Schlumberger, and Science Applications International, Inc. In 1995, Deutch became the Director of the Central Intelligence Agency (CIA).

While he was Under Secretary of Defense, he didn't want to spend *any* defense money on *any* new space launch initiatives, expendable or reusable.

So Denman felt safe in withholding the money.

Pressure from Congress had zero results. Just because an agency has funds to do something doesn't mean that those funds have to be used.

In the meantime, Col. Pete Worden left to work for Maj. Gen. W. Thomas West, special assistant to Air Force Chief of Staff Gen. Merrill McPeak. The reasons Worden left were not clear at the time in view of the Air Force reluctance to provide strong backing for the SSTO concept. In 1995, Council members learned that the numerous Air Force general officers couldn't openly support SSTO but didn't want the Air Force to be left out of the NASA program. The internal politics of the Air Force are such that its space advocates don't have the power of the fighter pilots who are running that service. Therefore, they have to support a series of technology development projects too small to draw attention.

Then came the real kicker: In a play where the ball went from Denman through Deutch to various Defense Department offices to Defense Secretary William J. Perry to the White House and thence to Congress, Denman tried to give the SSTO money back in spite of promises he'd made to the Air Force Chief of Staff that the funds would be released to continue DC-X testing.

Few people outside Washington realize that a fund recision process exists. So a brief primer is in order here.

Recisions or the cutting of monies or programs can be proposed by the Executive Branch to the Legislative Branch at any time (Title X, 1974 Impoundment Control Act). No reasons need be given. Recision proposals are usually bundled and submitted to Congress with the newly proposed budget for the next fiscal year although recisions can be proposed at any time. Once the Executive Branch proposes a recision, Congress has 45 "legislative days" to vote the recision into law or the recision ceases to exist. In the meantime, the funds for the recision items are put into escrow and can't be touched. If 45 legislative days pass and no action is taken, then the monies are removed from escrow and can be spent on the budget line items for which they were originally appropriated.

Or the funds can be "reprogrammed" from one specific account to another by the executive department involved. If the amounts are under $4 million, the reprogramming can be done without notifying Congress provided the reprogramming doesn't lead to zeroing-out the account or if the item is "not in the Congressional interest." If the amounts are between $4 million and $10 million, Congress must be notified, but congressional approval isn't required. From $10 million up to about two dirksens, reprogramming requires positive permission from each of the defense committees of Congress. All of this is done by means of letters.

Both the recision and the reprogramming procedures can be used to delay a project and thus essentially kill it. Denman and Deutch tried both. Some staffers on the Hill had to know it was coming, especially people like Dawson who would then provide the grease to slide the recision proposal through their respective committees.

On December 31, 1993, the Comptroller of the Department of Defense submitted recision proposals for the following programs: The Army's TOW anti-tank missile, the Navy's SH-60 helicopter, part of the Navy's shipbuilding and conversion program, the LAND-SAT project, and the "ARPA space program." The total proposed recision amounted to $314.7 million of which the "ARPA space program" was $50 million. Normally, Congress is delighted when an executive department says it doesn't need to spend some money.

Not this time!

When the SSTO supporters went to work on this, Denman changed his mind three times in three days about appealing the SSTO funding recision. Deutch kept his head low. But it was now obvious that the "ARPA Space Program" didn't get on the kill list by accident. It also was quite clear that the SSTO program faced more than just cautious foot-dragging. People were now out to kill it openly.

This had to be caused by active opposition. Rescinding $50 million out of a multi-billion dollar defense budget wasn't real parsimony. The SSTO supporters looked upon it as being pound wise and penny foolish.

The proposed SSTO recision didn't survive Congressional scrutiny because SSTO had found new supporters in the persons of then-House Minority Whip Newt Gingrich as well as John Murtha and Robert Walker. Everyone expected Congressman Dana Rohrabacher to be in the fray, and he was.

People are baffled that appointed civil servants with no accountability can oppose the will of elected officials in Congress. However, it's now apparent that the recision procedure is widely used. The "ARPA Space Program" recision attempt just brought it to the attention of many people who hadn't known of the procedure. When SSTO supporters discovered this ploy and killed this bureaucratic attempt to stop the program, it seemed unlikely that people in the Executive Branch would try to use it again. However, it was tried unsuccessfully several times over the next two years. And it did draw out from under the rocks some more of the SSTO opponents so they could be tagged and thereafter watched closely by the SSTO community.

The fight over the recision took nearly all of January, 1994. The recision proposal list was approved by Congress *with the exception* of the "ARPA Space Program." That portion of the recision list went back to the Department of Defense with letters from Congresspeople that said, in effect, "We gave you this $50 million to do a specific job. You wanted to give it back. We don't agree with you. You have the money. Now use it to do what we told you to do with it!"

If active opposition couldn't kill the SSTO program, foot-dragging still could, and people like Denman, Deutch, Dawson, and others knew it. The next thing they could do was to reprogram the funds into other projects.

Major Jess Sponable, now in total charge of the SSTO program, knew that under law he'd have to cancel the DC-X contract on February 1, 1994, unless the ARPA funds were released to him.

In the nick of time at 4:00 P.M. EST on Monday, January 31, 1994, the cavalry came over the hill and saved the DC-X.

The rescue mission appeared in the form of funding in the amount of $900,000 to keep the DC-X at White Sands and to cover the Clipper Site overhead charges until the ARPA funds could be released.

The money came from NASA Administrator Goldin.

SIXTEEN

Success and
Successful Failure

THE NASA MONEY WASN'T ENOUGH to get DC-X flight testing started again. Denman at the Advanced Research Projects Agency continued to sit on the funds although he'd been told in no uncertain terms that it was the desire of Congress that the money be released, that the DC-X flight tests be resumed at the earliest possible time, and that the Request for Proposal (RFP) be issued for the follow-on SSTO test vehicle.

In the meantime, bipartisan support for a government-funded SSTO X-vehicle program continued to grow. At the suggestion of Congressman Newt Gingrich, NASA Administrator Goldin initiated contact with Dr. Jerry E. Pournelle, chairman of the Citizen's Advisory Council on National Space Policy.

None of the growing Congressional support for X-vehicles seemed to faze Denman who stubbornly held his ground, apparently with the support of his boss, Dr. John M. Deutch. The new Secretary of Defense, William J. Perry, was pretty much out of the loop, being forced to deal with the problems of a Department of Defense that had just won a 45-year-old war with the Soviet Union. The Defense Department had to be "restructured" from the wartime footing it had maintained for nearly three generations. The people of the United States have always been suspicious of maintaining large standing military forces in times of peace. Figuring out how much of what kind of armed forces would be needed in the decades to come was a knotty problem.

In the minds of many defense planners and policy makers, the role of space forces in the proposed future mix was unknown. Certainly, space assets played a major role in the Cold War and especially the Gulf War. Peacetime utility ("dual-use technology") was already in place for such military space assets as meteorological and communications satellites. The Global Positioning System (GPS) was quickly finding low-cost commercial uses.

What was the role of SSTOs, reusable launch vehicles, expendable launch vehicles, and indeed space activities themselves in the emerging "new world order?"

One of the major efforts during early 1994 was a study mandated by the FY1994 Defense Authorization Act. It directed the Secretary of Defense to "(a) develop a plan and establish priorities, goals, and milestones for space launch modernization for DoD or, if appropriate, the government as a whole, (b) allocate funds to do this for ARPA, (c) identify new launch system requirements (if required) and pursue innovative government and industry funding, management, and acquisition strategies, (d) define cost reductions for current launch vehicles, and (e) study differences between U.S. and foreign space launch systems." This study was conducted by a team led by Lt. Gen. Thomas S. Moorman, Jr., of the Air Force.

The "Moorman Report" quickly became an excuse at ARPA for doing nothing. Denman stubbornly refused to release the $50 million for continued flight tests and follow-on design studies until the Moorman Report was completed, lest the money be given in advance to a program that didn't have the blessing of this study. Thus, the logic went, ARPA could be faulted for spending the money on a program that didn't fit into the future defense plans. On the other hand, if the SSTO program was part of the study recommendations, the expenditure of the funds would be justified.

The Office of Science and Technology Policy (OSTP) of the White House came out with a draft of a proposed national space policy in which NASA would be responsible for reusable launch vehicles. Some of the language in this OSTP report was fought strenuously by pro-SSTO X-vehicle advocates on one side and the status quo factions led by people such as Dawson on the other.

The original policy would have perpetuated the "business as usual" space program as a government monopoly. SSTO supporters wanted to see the government do fast-track experimental vehicles to lower the perceived technical risk, thus inducing the private production of spaceships that could be sold to space transportation operators ("spacelines") and even to government users such as the Department of Defense. The SSTO supporters saw this as comparable to the Kelly Act that authorized the government to give airmail contracts to the fledgling airlines of the United States in 1925.

While this was going on, an agreement was made between NASA and the Ballistic Missile Defense Organization concerning the DC-X. NASA Administrator Goldin wanted to take over the DC-X and have it shipped back to Huntington Beach, California, where McDonnell Douglas would replace the heavy aluminum tanks with lightweight graphite-epoxy and aluminum-lithium ones. Other improved systems would be added in this upgrade to produce the DC-XA. Therefore, BMDO agreed to turn over the DC-X to NASA following the completion of the flight test series.

People at NASA Headquarters and George C. Marshall Space Flight Center in Huntsville were beginning to plan the follow-on SSTO X-vehicle that was referred to as an ATD (Advanced Technology Demonstrator). It would be a three-year program costing about $300 million. The vehicle would be roughly twice the size of the DC-X, would use lightweight composite structures, would use up to eight RL-10 hydrogen-oxygen engines, and be capable of attaining a suborbital altitude of 600,000 feet (about 113 miles). Attempts to make it into an operational reusable sounding rocket for scientific research were successfully countered by careful work by Council members because the vehicle was intended to be an experimental vehicle (X-vehicle) with no purpose other than to build, fly, and test some of the remaining technologies necessary for a full SSTO. It was seen as a cooperative government-industry effort with innovative cost-sharing, market development, and guarantee arrangements. This concept gave birth to a sort of new institution called "anchor tenancy" where the federal government would guarantee certain use levels. This was looked upon as equivalent to the historical precedents of the railroads and the airlines, but it wasn't.

Many of the members of the Citizens Advisory Council didn't believe that the follow-on vehicle was necessary and, if it was to be built and flown at all, it should be capable of "scaring the hell out of orbit" or even going into orbit—with or without a payload—and returning.

In the meantime, Sponable and Gaubatz sweated out the funding delay. This was extremely frustrating because Congress had approved and authorized the money, given it to ARPA, overridden the recision request, and told ARPA to spend the money as Congress had directed. ARPA stubbornly continued to sit on the flight test funds while the DC-X sat in its canvas hangar amid the temperature extremes and blowing sand of Clipper Site. Thanks to the NASA money, the Delta Clipper Team was held together. But it couldn't go on this way without the ARPA funding.

The "Findings and Recommendations" portion of the Moorman Report became available on April 15, 1994.

One of these findings said that "there exists general consensus on the potential benefits of a new reusable system but there are widely divergent views on timing, approach, cost, and risk." The report recommended that the Defense Department and NASA pursue a technology maturation effort that would include experimental flight demonstrations.

Yet another finding pointed out that Defense and NASA space launch program coordination needed improvement. This led to the recommendation: "Assign the Department of Defense the lead role in expendable launch vehicles and NASA the lead in reusables."

The Moorman Report could no longer be used as an excuse for delaying the funds. Another cliff-hanger date dawned: April 30, 1994. At that point, the NASA money would run out.

And at 4:00 P.M. on Friday, April 29, 1994, Major Jess Sponable was informed that $5 million of ARPA funding had been released to continue the DC-X flight tests.

But none of the additional money was released for the follow-on. If NASA was going to do reusable launch vehicles, Denman wanted to keep that money in ARPA for small technology feasibility studies, not for building and flying hardware. Denman was playing it safe. If you don't build and test hardware, you won't have any failures to explain. If you produce paper reports, a computer analysis can always show that things will work perfectly.

Meanwhile, Sponable and Gaubatz had to get the DC-X flying again after a delay of more than half a year.

Paul Klevatt's Delta Clipper Team hadn't been idle. They'd closely monitored and inspected the DC-X during those long months, gathering data on what happens when a reusable spaceship isn't being reused. The DC-X turned out to be a very robust rocket. The stand-down had only caused a few seals and gaskets to dry out. These were easily replaced, thanks to the fact that the DC-X had been designed for easy maintenance. Everything else checked out in the green.

But it took about six weeks before the DC-X flew again.

It did so with great success at 8:42 A.M. MDT on Monday, June 20, 1994, at the White Sands Clipper Site.

The purpose of the flight was to further "expand the performance envelope" just as test pilots push an airplane higher and faster to get actual data on the performance.

The DC-X lifted off and climbed to a height of 1,500 feet above the launch site. It then followed a curved ascent to 2,600 feet, travelling laterally 1,050 feet sideways. The flight controls then told the DC-X to reverse its direction of flight and climb to an altitude of 2,850 feet. During this flight profile, the ship went through an aerodynamic angle of attack—the angle between the longitudinal axis of the rocket and the flight path—from zero to 70 degrees. Once over its landing site, it descended vertically and touched down 136 seconds after liftoff. "The Little Rocket That Could" proved again that it still could after eight months of sitting in mothballs.

And it proved it again on Monday, June 27, 1994, during the fifth flight. This was planned to be similar to the previous flight except to push to higher altitudes and airspeeds.

The flight was scheduled for 9:00 A.M. MDT. We'd rolled out of bed in Las Cruces at 4:00 A.M. in order to get to White Sands Missile Range 30 miles to the east. There, we were loaded on the usual Army school buses with "natural air conditioning" (just open the windows) and bussed 30 miles north to Clipper Site. It was a far smaller crowd than in September 1993, about 100 people altogether. I received press credentials because I was writing this book. Several others were there—Max Hunter, Bill Gaubatz, Henry Vanderbilt, and a contingent from NASA Marshall Space Flight Center that would take over the DC-X after the test flight series was completed.

The flight preparations—as before, it wasn't a true count down but more like getting an airplane ready for departure—were completed early and range safety clearance was obtained for launch, which took place at 8:37 A.M.

We were used to seeing the bright flash of hydrogen vapor deflagration at engine ignition. However, this time a much more intense and bright orange ball of fire appeared on the south side of the DC-X.

Videotape made by Henry Vanderbilt and White Sands—the former was immediately available for study, the latter showed up a few weeks later—clearly showed an external, not internal, hydrogen explosion. In normal operation, a cloud of free-vented gaseous oxygen, gaseous hydrogen, and water vapor is created by the engine cool-down procedure prior to ignition. This cloud usually gathers around the base of the DC-X and rises. It's further dissipated by air blown from an air purge duct in the pad. This duct is 12 inches in diameter and 25 feet long. Air is continuously blown across the launch pad by a blower to disperse this cloud of gases.

This time, the wind blew the cloud so it was ingested into the duct by the blowers. When the RL-10 engines ignited, the gaseous mixture in the duct detonated.

This caused the remainder of the cloud to detonate, creating a pressure wave that in turn blew in the flat south side of the DC-X graphite-epoxy aeroshell. Then the pressure wave was reversed, pulling the aeroshell structure outward, causing it to bulge and tear.

As the DC-X rose, it began shedding pieces of the aeroshell from its south side. I saw several big ones fall off as the DC-X climbed skyward.

Flight director Pete Conrad didn't see the pieces coming off. He was inside the Flight Operations Control Center trailer watching the telemetry data from the DC-X as it was displayed on his computer screen.

Sponable threw open the door and yelled, "Pete, abort!"

"Why? It's flying fine!" Conrad replied, looking at the computer screen. And, indeed, the DC-X was flying perfectly on its assigned flight path.

"Pete, abort! We can see pieces falling off!" Sponable told him.

Conrad moved his computer mouse and triggered the autoland command on the screen.

Autoland is an emergency command sent to the DC-X that tells the rocket to land at once. At 17.7 seconds into the flight and an altitude of 1,100 feet, the DC-X landing legs popped out. Conrad's voice on the PA system called out, "Autoland!"

The DC-X climbed another 1,500 feet on momentum with aeroshell pieces continuing to fall off. Then, as it started down, the pieces seemed to be falling up and away from it.

The DC-X landed amidst a cloud of white gypsum sand 77.9 seconds after liftoff. It hadn't landed on the concrete pad but about 200 feet to the southwest on the white sands of the desert.

"Weight on gear! Engine shut down!" came the call.

As the cloud cleared away, we saw the DC-X was still upright.

We'd witnessed the first emergency landing of an experimental spaceship.

If the DC-X had been an expendable rocket or even the space shuttle, it would have been a smoking hole in the desert instead of standing forlornly on the white sands with one quarter of its aeroshell ripped away in a 4-foot by 15-foot gash.

It was an experimental flight that Sponable, Conrad, and Klevatt hadn't expected to make. But it proved that a reusable spaceship could be built with an abort capability.

As such, the flight was an enormous success. We learn a lot from failures, and this was certainly the case here.

The DC-X still had nearly half its load of liquid hydrogen and liquid oxygen in its tanks. So no one but the ground crew was allowed down to the launch and landing site. We climbed aboard the naturally air conditioned buses and went back to the base.

The following morning as Henry Vanderbilt and I were preparing to check out of the hotel in Las Cruces and go back to Phoenix, the telephone rang. It was Debbie Bingham of the White Sands Public

Affairs Office. "Colonel Sponable says it's okay for you to go up to Clipper Site and view the DC-X. The bus leaves in thirty minutes. Can you get here?"

I had driven U.S. 70 from Las Cruces to White Sands many times in the five years I worked there. Could we make it in 30 minutes? Yes, but we flew low.

It turned out that we were only two of the four people who hadn't gone home and thus got to Clipper Site to see what had happened the day before.

It looked worse than it actually was. The DC-X had been put back on its launch stand with a crane. All of the aeroshell pieces had been picked up from the desert floor. "Go anywhere you want," Paul Klevatt told us, "but please don't stick your head inside the aeroshell at this time."

The DC-X aeroshell looked like the fender on a Chevy Corvette that had been torn up by the disintegration of a steelbelted radial tire at high speed. Nothing that a little Bond-O and duct tape wouldn't fix, I figured. (Actually, since the aeroshell is more than mere fiberglass, the graphite-epoxy material would require a bit more in the way of repairs.)

We went out to where the DC-X had landed in the gypsum sand. The rocket motors had dug four holes about four feet in diameter and a foot deep. "Another UFO landing site," someone quipped.

The "press conference" was held in the mobile home that served as a combined office and storeroom for the Delta Clipper Team. It was more of a bull session with only five of us there— Henry Vanderbilt of the Space Access Society, the reporter from the local Alamogordo newspaper, the publisher of a space advocacy newsletter, Sponable, and me.

"Burt Rutan of Scaled Composites, builder of the aeroshell, will be here Friday to assess the damage," Sponable reported. "There's a minor crack in the hydrogen tank caused by the south flap actuator rod being jammed into it by the explosion. Other than that, the bird is in good shape. We won't know the various options for repairing it until we assess the damage a little more. We'll certainly fix the problem with the air duct! However, the fast three-day turnaround between flights obviously is out of the picture right now. We can't

FIGURE 16-1: *An external hydrogen explosion tore the south side of the DC-X graphite-epoxy aeroshell at ignition during the June 27, 1994, flight test. In spite of this damage, the rugged DC-X performed a successful abort and landed. The author was one of four people who got to the launch site the following day. (Photo by G. Harry Stine.)*

FIGURE 16-2: *"Must be another UFO landing!" Paul Klevatt and Lt. Col. Jess Sponable view the four small potholes dug in the desert sand by the rocket engine exhausts when the DC-X landed following the damage and abort on June 27, 1994. (Photo by G. Harry Stine.)*

FIGURE 16-3: *"We'll fix it and we'll fly it again," Lt. Col. Jess M. Sponable tells us at the "press conference" held at Clipper Site on the morning of June 28, 1994. (Photo by G. Harry Stine.)*

fly again on Thursday. It may take several weeks to repair the bird and check it out carefully for any additional damage."

This is what flying experimental vehicles is all about. Build a little, fly a little, patch a little.

But some people in Washington still didn't catch on.

I was afraid that some NASA people would quietly say to each other, "If this is the way BMDO runs a rocket program, we'll show them how to do it right! We'll make sure we have all the problems solved before we commit to a flight! We'll never permit a failure like this to happen!"

Indeed, that was exactly what some of them were thinking, if they didn't say it. They didn't understand that the June 27, 1994, DC-X abort was the sort of occurrence that one should expect in an experimental program.

SEVENTEEN

Sponable's Rules and the X-Vehicle Hearing

RECENTLY PROMOTED, Lt. Col. Jess M. Sponable decided that the only logical thing to do was to fix the DC-X, fly it again, and then turn it over to NASA. But his superior officers and civilian managers at Advanced Research Projects Agency (ARPA) didn't agree at first. Burned by the failures of the previous space vehicles they'd funded, they exhibited a surprising lack of testicular fortitude as a result of the DC-X explosion and abort. Initial letters to Sponable told him to cancel the whole program. He responded with a rational and reasoned reply, including comparative costs of canceling versus repairing. NASA also reminded ARPA that their agreement required the DC-X to be turned over to NASA in flightworthy condition. NASA strongly suggested, since ARPA already was sitting on the funds, that ARPA approve the repair of the DC-X and the completion of the flight test series.

The DC-X was trucked back to Huntington Beach for repairs, and the flight tests were scheduled to resume whenever the aeroshell was fixed.

Sponable had learned a lot during his tour of duty as Project Officer for the X-30 National AeroSpace Plane (NASP) and in the project office for the SSTO program. In the interval between those two positions, he'd attended the Air Command and Staff College at Maxwell Air Force Base. Sponable is a career USAF officer who wants to get his service and the rest of the nation into space economically, reliably, dependably, cheaply, and on-demand. He has vision, but he's never allowed that to cloud his view of reality.

During his stint in those cramped little Pentagon basement cubicles of the SSTO project office, Sponable codified the ten basic "fast track" management principles that were developed for the Ballistic Missile Defense Organization programs such as Delta 180 (14 months from proposal to launch), Delta 181 (18 months), Delta Star (13 months), LOSAT (13 months), MSTI-1 (12 months), Clementine (23 months), and DC-X (22 months).

All of these are applicable in the modern, high-tech corporate environment as well.

1. Agree to clearly defined program objectives in advance. Establish a vision. Push the envelope. Create a sense of urgency. Don't settle for incremental advances. Ignore the criticism and press ahead. Agree to cost, schedule, and top level functional objectives versus detailed requirements. And avoid detailed specifications. All of these things require agreements between the program manager and the program executive officer and between the contractor and the customer. The program manager should zealously guard against changing requirements. As Lt. Gen. George S. Patton advised, "Never tell people how to do things. Tell them what to do, and they will surprise you with their ingenuity."

2. Appoint a single manager under one agency or boss. Empower the program manager to make all decisions consistent with law or established organizational policy (then don't change the policy). The program manager should report directly to one boss. Funds should have no strings attached. The program manager should understand technology and system trade offs, should implement the program as an honest broker, should consider program success more important than the next promotion, and be able to make decisions, often quickly and with limited data. The military services call this "unity of command."

3. The program offices—those of both the contractor and the customer—should be kept small with a short "tooth to tail" ratio. All team functions, technologies, and disciplines should be colocated and organized into concurrent engineering teams that develop a sense of "ownership." Design for quality, operability, and supportability up front, not after the fact. Push responsibility to the lowest possible level. The program manager should be free to hand-pick the critical managers, and all managers should be working managers. During the SSRT program, the government program office in the basement cubicle of the Pentagon consisted of one to two people while the MDC office had between five and fifteen people. The whole MDC team consisted of about 100 people. "Business as should be" requires 10% to 50% the number of people as "business as usual."

4. Build hardware, not paper. Sponable points out that "one prototype is worth a million words." Keep the contractor competition sharp, and select quickly for quality. Build quickly. Don't forget software, so freeze code modules early. Set demanding schedules and push hard. Many years ago, the Packard Commission report on the organization and operation of the Department of Defense stated, "A high priority should be given to building and testing prototype systems and subsystems before proceeding with full-scale development. This early phase of research and development should employ extensive informal competition and use streamlined procurement practices."

5. Focus on key demonstrations of experimental versions and don't worry if these fail. The team gets the credit for success, and the program manager takes any blame for failure. Build a little and test a little. Don't try to demonstrate everything at once.

6. Streamline the required documentation and reviews. Use a simple statement of work and reduce the required review data and mileposts. Empower the contracting officer to act autonomously without major executive review of every step. Hold reviews and meetings only when decision-makers are present. Don't micromanage, but ensure that decision logic is sound. Resolve open issues rapidly. Minimize the number of drawings and internal documents. Use the shop floor as part of the concurrent engineering team. And use "best commercial practices."

7. Use the contractor, not the customer, as the integration and testing operator. The contractor who builds it knows it, and this reduces customer costs and delays. It also shrinks the test schedule, saving time and money. Let the contractor retain "ownership" to simplify any liability issues and to build esprit de corps. Build the X-version to test and get data for the next version and retire the old X-version as soon as objectives are achieved. And don't build a testing empire.

8. Develop minimum realistic funding profiles consistent with program objectives and schedule. Provide adequate funds but maintain customer control of funds distribution. Don't give out too much money too soon, but keep in mind that front loading is essential when building hardware. Relentlessly seek a stable funding source. Track life cycle costs, not just development costs. The contractor program manager must personally approve all costs in advance or in real time and must share all this information with the customer program manager.

9. Track the costs and schedule in real time, using the contractor's in-house cost tracking system. Keep cost accounts to a minimum and focus on controlling risk. Minimize surprises by timely and accurate forecasting. Put a first-rate business team in place to react rapidly to cost and schedule problems and to avoid them if possible. Timely and accurate data, as well as excellent communications, are required between contractor and customer to solve anticipated problems and surprises.

10. Finally, mutual trust is essential. The contractor and customer program managers should share all program information, evaluate the risks, and make the critical decisions. The customer must not second-guess every contractor decision and must trust the contractor to build, fly, and test the hardware. The contractor must trust the customer to provide stable program direction and not to drive the costs and schedules with unreasonable demands. People need managers, and managers must be there for the team. So think team, team, team and build an esprit de corps that says, "Anyone can do it, but we *do it!*"

Sponable insists that a good leader must provide personal incentive, incentive, incentive, because personal incentive and the *kind* of incentive were two of the things that gave the United States a distinct advantage over the Soviet Union.

It was fascinating to watch Sponable and Gaubatz put these principles into operation on the DC-X program.

NASA Administrator Daniel Goldin seemed to be catching on about what was needed to do this sort of fast track development. Goldin's self-imposed goal was to restructure and rebuild a space agency that wasn't enjoying the highest reputation among people on the Hill, much less the American taxpayer. He admitted that he didn't know exactly how to do it, so he listened to what people told him. He asked for and got inputs from his own NASA people. Some of what the aerospace industry executives and managers told him was discounted. Goldin was most interested in the advice of people who had no vested interest. He was talking to Dr. Jerry E. Pournelle and getting additional input from Congresspersons and staffers as well.

However, if Goldin caught on to the fast track policies that Sponable used, the concept seemed to go right past a lot of NASA's troops. When old-line career NASA and aerospace industry types heard "quick and dirty SSTO demonstrator," they still thought of 36 dirksens and 20 years for developing Shuttle II. To other NASA people, it meant a dirksen and five years rather than what was really said: $300 million dollars and three years. "High flight rate" meant flying several times a year off a rebuilt Saturn pad at the Cape with several thousand people instead of from a concrete pad in the desert with no more than 30 people. Risk-taking was something one didn't even think about because failure hadn't been tolerated in NASA . . . and jobs were hard to find on the outside.

It wasn't clear to SSTO supporters that NASA could indeed get on this fast track. Many feared that giving SSTO to NASA meant that no one would ever see any hardware flying in this century, only more studies.

Others were worried that NASA would turn the SSTO into another jobs program. NASA Administrator Goldin flatly stated that this wouldn't happen (at least on his watch). He first went on public record concerning this during a luncheon speech at the National Press Club on Monday, June 20, 1994, when he answered a question about funding new man-rated spacecraft: "If we go to a new machine . . . it is not going to be the same old way. I don't think NASA's going to go back to Congress and say, 'We need billions of dollars so we can go make a big program to try and get a replacement for the shuttle. . . .'"

In responding to a question about his support of SSTO, he replied, "Delta Clipper is one of the most imaginative programs that this nation came up with for access to space, and the Ballistic Missile Defense Organization was shepherding this program until they ran out of money. I will say that NASA put its money where its mouth is and this Administrator came up with money to save that program. We are enthusiastic about it, and now the Department of Defense is going to perform a number of tests to complete the test series on the Delta Clipper to prove the feasibility of the model they built. And we're working with the Department of Defense to come up with a program that could help make this technology a reality and cut the cost of access to space and make the reliability much higher."

Someone in NASA apparently wasn't listening to what the boss was saying in public. Later that very week, NASA made 16 nonprofit cooperative agreements with aerospace companies to develop a broad range of reusable launch vehicle technologies leading to the goal of *choosing* a new reusable launch vehicle by the year 2000. NASA didn't use the SSTO term but substituted their own: Reusable Launch Vehicle (RLV). These awards came out of Marshall Space Flight Center in Huntsville, Alabama, whose people seemed wedded to the ammunition paradigm they'd learned from the old German rocket team.

What NASA Marshall Space Flight Center wanted was complex. Several engineering teams were conducting in-depth studies of space launch vehicles under the NASA Access to Space project. One of the conclusions was to let several industry teams *design* (not build) prototype SSTOs in the 1995–1996 time frame. This meant more paper studies. One contractor would then be selected to build a 100,000-pound suborbital SSTO in 1997–1999 leading to a fully operational orbital RLV in 2000.

The 16 agreements covered a broad range of technology developments with the idea of getting all the technology in hand to build the new NASA reusable spaceship in the year 2000. This sounded like a verbatim repeat of the approach that NASA people had advocated over the preceding year.

The boss was saying one thing, and his troops were giving away money to do something else.

Max Hunter and I discussed this when we were together for the June 27, 1994, flight at White Sands. Calling upon his decades of experience both in the government and the aerospace industry, Max believed the NASA approach would produce Shuttle II, a spaceship owned and operated by the government.

At that time, most NASA manned spaceflight people wanted Shuttle II because they could see no other way to do it.

By mid-1994, it became obvious to many former fence-sitters and detractors that a reusable SSTO was the way to go. But between the Big Perfect Spaceship people and the X-vehicle advocates, there was no agreement on how to proceed.

This became clear when the confrontation between the staffers of the House Science, Space, and Technology Committee and members of Congress and their staffers could no longer be delayed.

Dawson and his colleagues wanted to build the perfect SSTO. Also, it was beginning to look like this was the path that NASA was taking because it was consistent with their no-failure policy.

On the other side of the fence, SSTO supporters wanted NASA to build and fly X-vehicles.

On July 15, 1994, Dawson agreed to hold a hearing on reallocation of SSTO funds. This was approved by Committee Chairman Ralph M. Hall. The hearing had to be held quickly because the mark-up of the FY95 NASA Authorization Bill was scheduled for Wednesday, July 20, 1994. So the hearing was scheduled for Tuesday, July 19, 1994.

In opening remarks, Congressman Dana Rohrabacher summed up the purpose of the hearing as follows: "NASA this year for the first time requested funding from Congress to conduct SSTO research. Over the past year, NASA has concluded that SSTO is technically feasible as well as being reasonable and practical. Its Access to Space study found that SSTO will save on launch costs and will provide reliable and cheap access to space. This hearing, then, is about the proper way to proceed with this valuable program. I maintain the best method is the way we as a nation *used* to develop advanced systems. . . . The construction of X-vehicles is as old a development strategy as powered flight. The Wright brothers built experimental gliders in 1901 and 1902. They flew them in order to obtain empirical flight data and experience so they could then build the 1903 *Flyer*—arguably the *first* X-aircraft."

The lead witness was Thomas J. "Jack" Lee, then Special Assistant for Access to Space in NASA Headquarters, who presented the NASA position and, knowing that many of the Committee members were strongly in favor of the X-vehicle approach, centered his remarks around descriptions of the three test vehicles NASA wanted to build and fly—(1) the DC-XA, (2) a Small Booster Technology Demonstrator (SBTD) for testing propulsion systems and advanced thermal protection systems, and (3) an Advanced Technology Demonstrator that would be an unmanned, suborbital, fully reusable vehicle. No one knew at that time where the Small Booster Technology Demonstrator came from or what function it would perform in the program. *Lee gave no projected schedule for doing all of this.* Furthermore, NASA appeared not to be wedded to the SSTO concept because he mentioned Two-Stage-To-Orbit on an equal footing with SSTO. He also stated NASA's intent to do considerable ground testing on several technologies NASA believed were not yet ready for application. These technologies are (the parenthetical additions are my comments):

Graphite Composite Primary Structure. (Boeing uses a lot of graphite composite materials in the 747-400, 757, 767, and 777 airliners where whole surfaces such as the horizontal stabilizers are made from this material. FAA has certificated these airliners. People are riding in composite airliner structures.)

Reusable Cryogenic Tanks. (USAF has done this and the reports of the USAF Have Region and X-30 NASP research show this.)

Long Life/Low Maintenance Thermal Protection System. (McDonnell Douglas says they've already got it.)

Vehicle Health Management and Monitoring. (Military aircraft such as the F-15, the F/A-18, and the F-22 already have it.)

Autonomous Flight Control. (Fully automatic autopilots and other flight control systems have been flying for 25 years in the Boeing 747, Lockheed L-1011 TriStar, and Douglas DC-10.)

Operations Enhancement Technologies. (By this, NASA probably meant "operability, supportability, and maintainability" that has already been tested in the DC-X. It would help if everyone involved in the SSTO matter spoke the same language.)

Advanced Propulsion Systems. (The availability of suitable engines has always been a problem in aerospace design. Engineers always want a lighter, more powerful, and more efficient engine, be it reciprocating, jet, or rocket. But this never held back aeronautical progress because engineers designed using available engines and later upgraded the vehicles to take better engines. Witness the Douglas DC-4/DC-6/DC-7 series and the Lockheed Constellation propeller-driven airliners. The list includes more recent designs such as the Boeing 707 and Douglas DC-8 jet airliners. SSTO is do-able with existing engines such as the old J-2 that powered the second and third stages of the ancient Saturn V moon rocket. The development of a plug nozzle or aerospike rocket engine would only improve what can already be accomplished. Holding back SSTO for lack of a perfect engine doesn't make good sense.)

Dr. Ivan Bekey presented a long and highly detailed technical testimony on SSTO technology, concluding that NASA should test small-scale models to develop a data base.

Dr. Jerry Grey, Director of Aerospace and Science Policy for the American Institute of Aeronautics and Astronautics (AIAA) and a visiting professor at Princeton, strongly favored ground testing over flight testing. He stated that the DC-X flight did not test any new technologies. (Either he hadn't done his homework or he doesn't understand what technology is all about. The DC-X was built and flown with existing technology to show what could be done.)

Bill Gaubatz of McDonnell Douglas wanted to build and fly X-vehicles. Although he stressed proper management, he also stated, "We strongly believe that X-vehicles and their supporting systems must provide the focus for development and testing of all components. . . . Flight tests must evaluate operability and supportability under real world conditions. *The technologies needed to realize an operational SSTO are mostly in hand.* [My emphasis.] To be cost effective, all SSTO development must be conducted in a rapid prototyping environment." Gaubatz knows these things; he and his team "bent tin" and made it fly.

Col. Pete Worden came out flatly in favor of a strong X-vehicle program. "Generic technology development is no substitute for a flight program. Without a flight program there is nothing to focus technology development. Without the flight experiments, technology programs quickly decay into 'sandbox' activities which survive more on their contribution to local employment and vested interests than on real future needs. . . . *Unless [a] program is centered on a flight demonstration within a few years—I would suggest no more than three—it is my opinion that it is pointless to proceed."* [Colonel Worden's emphasis.]

Daniel O. Graham, Executive Director of the Space Transportation Association, stated, "I know of no enthusiast for SSTO who would maintain that no ground testing of components is necessary. Nor do I know any who would not agree that rocket engine development must be pursued if the maximum potential of fully reusable space transport systems is to be realized. On the other hand, I know of no engine experts who maintain that SSTO cannot reach orbit using existing engines. . . ."

Then Graham put the real long-term issue in focus. "The prospect of profitable private launch enterprises using SSTO-type vehicles has already spurred action in the commercial field. At least three entrepreneurial groups are actively seeking private financing for space launch companies operating fully reusable vehicles. Whether or not these efforts are successful, they represent recognition of fiscal reality and confidence that launch costs can be sharply reduced. . . . The Space Transportation Association in our white paper, 'Space Policy 2000,' the Aerospace R&D Policy Committee of the Institute of Electrical and Electronic Engineers, and the six aerospace company Alliance have all concluded that the commercial space market can expand dramatically if space transportation costs are cut dramatically, thus encouraging private investment."

If there could be winners and losers in such a debate, it's obvious that the X-vehicle contingent won. Note that there was no dissent among the witnesses about the feasibility of SSTO, and this was a complete change from the situation only a year previously.

An unusual "Additional Views" statement from the Committee was prepared and signed by 10 of the 12 Committee members who were present for the hearings. It says:

> We believe . . . that the best way to achieve a demonstration of SSTO capability is through the rapid development and construction of flight hardware to test those technologies currently in hand and those technologies that have been rapidly developed for SSTO. . . . The key to future SSTO development is to build SSTO X-vehicle demonstrators that fly higher and faster and have a quick turnaround. These vehicles should be built quickly on a step-by-step approach that learns from the flights of the previous X-vehicles. We support new technology developments but do not think that demonstration of X-vehicles should be delayed as we wait for the maturation of these new technologies . . . X-vehicles should explore *what we can do with what we currently have* as well as evolve with the advanced in technology. . . . We remain concerned that the X-vehicle concept will be lost in the excessive analysis that typically occurs at NASA which results in designing and testing the concept to death before something is actually built and flown. It is all too easy to put so many people on a project that no system can satisfy all their demands and concerns, and thus the project remains a paper study with no flying hardware; or results in a design that is so expensive that you don't dare test its limits. . . . Alas, this is a well known cultural bias at NASA . . . and overcoming it will require constant vigilance by NASA as well as by Congress. To succeed, NASA must copy the type of private sector operations that produce innovation and inventiveness.

NASA Administrator Goldin had decided to try this.

In retrospect, the hearing—given little attention by the news media in common with most space related activities in the wake of the Cold War's ending—was a watershed event. Max Hunter's "riot" immediately grew in intensity.

EIGHTEEN

Space Policy, Code X, and the Round Table

THE PERIOD FROM MID-1994 through the end of 1995 revealed that the space transportation paradigm had started to shift. Hardly anyone, not even those deeply engrossed in making it happen, fully realized that this was taking place, much less the extent of it. The American public didn't know about it because the national news media ignored it, playing out the usual well-rehearsed coverage every time a space shuttle was launched. Given all the revolutionary changes taking place in the federal government and given the spectacular news stories dealing with murder trials and bombings, people in the television, radio, newspaper, publishing, and periodical industries went for the "news" and considered that "space was dead" because, after all, didn't NASA get defunded along with all the other pork projects?

Space was far from dead. However, the *nature* of space activities was changing. A recounting of important events that occurred during this period clearly shows an acceleration of activity in the direction of true space access.

The basic foundation for future space activities came on August 5, 1994, with the release of the new United States Space Transportation Policy by the White House. The consequences of this haven't been grasped yet by many people, even if they knew about it. It's mere existence comes as a surprise to most.

Prepared by the Office of Science and Technology Policy (OSTP) of the White House and signed by President Clinton, the new document showed that it hadn't been developed in a vacuum. The ideas of the SSTO advocates run through the text, which, although it makes policy out of many of the concepts presented thus far, still contains shortcomings and paradigms left over from the old rocket ammunition philosophies.

The 1994 United States Space Transportation Policy requires that the United States government promote the reduction in the cost of current space transportation systems while improving their reliability, operability, responsiveness, and safety.

It gives the Department of Defense the responsibility for improving and evolving the current expendable launch vehicle fleet.

NASA is charged with (a) improving and continuing to operate the space shuttle system, and (b) developing and demonstrating the "next generation reusable space transportation systems, such as the single-stage-to-orbit concept."

The new Policy covers three separate space transportation domains:

1. National Security Space Transportation relating to matters such as surveillance, communications, and other national defense factors.
2. Civil Space Transportation, the nonmilitary space program primarily run by NASA as a result of President Eisenhower's initial 1957 space policy separating military and nonmilitary space activities.
3. Commercial Space Transportation carried on between companies that operate space launch vehicles and organizations that purchase and use such services.

The Department of Defense is to maintain the Titan IV launch system until a replacement is available and may use the NASA space shuttle to meet national security needs.

NASA is to continue to operate the existing space shuttle fleet until a space shuttle replacement is available. No additional space shuttle Orbiters are planned. NASA is also to conduct technology development and demonstrations for the next generation SSTO spaceships.

The federal government is to purchase commercially available United States space transportation products and services to the fullest extent possible and "shall not conduct activities with commercial applications that preclude or deter commercial space activities, except for national security or public safety reasons."

The new United States Space Transportation Policy has a lot more to it, of course. But these are the salient points that bear upon the design, construction, and operation of spaceships by commercial companies for profit-making purposes. It opens the door enough to allow commercial space transportation to get under way without the one factor that has held back investors to date: the threat of government favoritism, obstructionism, and competition that history, as presented in this book, clearly shows has been a real detriment to private enterprise in space.

NASA Administrator Goldin continued to pursue the course of action he spoke of in public. On September 6, 1994, he appointed Dr. John E. "Jack" Mansfield as Associate Administrator for Space Access and Technology to head up a new office, Code X, at NASA Headquarters. Mansfield's responsibilities included the development of reusable SSTO spaceships.

Mansfield was well known and highly respected in Washington. He'd been chief scientist for the Senate Armed Services Committee when Goldin tapped him for Code X. As mentioned earlier, Mansfield was one of the authors of the mid-1993 Space Launch Oversight Trip Report.

SSTO now had an official at NASA Headquarters who was in direct charge and who reported to Goldin. Furthermore, he was a space advocate.

A little over a week later on September 14, 1994, the National Space Society sponsored a National Space Transportation Round Table for Reusable Launch Systems with the direct participation of NASA. The National Space Society is a nonprofit organization of space advocates that started out as the L5 Society and the National Space Institute founded by Dr. Wernher von Braun. Although individuals who sat at the table had to be carefully chosen because so many people wanted to be there, the meeting held at the Marriott Hotel in Washington, D.C., was about as exclusive as a rainstorm.

Completing that simile, it was one of the watershed events in the affairs of SSTO. It appeared that the waters were all running in one direction: SSTO.

Fifteen direct participants gathered around the table and the video cameras started to run. NASA Administrator Dan Goldin began by stating, "We need a revolution." Incremental reductions of space transportation costs and complexities by only a few percent were unacceptable, he said. He wanted at least one order of magnitude reduction and preferably two.

Lionel S. "Skip" Johns from the White House staff told the participants, "Current and future space markets represent an engine of growth for the nation. Low cost, reusable space transportation vehicles are the keys to that engine of growth."

Jimmy Hill, Principal Deputy Assistant to the Secretary of the Air Force from the Pentagon said that the recent national space policy did not rule out Defense Department participation in the NASA reusable launch vehicle program and that Defense would support it "should the technology evolve" to the point where it became useful.

The aerospace industry brought in the heavies—Chairman and CEO John S. McDonnell of McDonnell Douglas, and Rockwell Chief Operating Officer Kent M. Black among them.

Also present from the Hill were Senator Howell Heflin (D, AL), George Brown (D, CA), and Dana Rohrabacher (R, CA).

But this wasn't a love-fest of government agencies and aerospace companies. Investors and venture capitalists such as Wolfgang Demisch of Bankers Trust Securities were at the table along with present and future users of space transportation services.

Ed Tuck, Vice Chairman of the Teledesic Corporation that intends to launch nearly 1,000 satellites by the year 2000 and another 1,000 in the opening years of the 21st century, drew attention to the fact that Teledesic was only one of nearly a dozen companies who want to launch constellations of communications satellites. His message was that if the United States wants to capture this market, it must have vehicles available before the year 2000.

Venture capitalist Shelley Harrison made a prophetic comment, "On one side of the table we have the government who needs a service, on the other side of the table we have industry who wants to provide a service, and down at this end of the table we have some people from Wall Street and venture capitalists . . ." Indeed, this was the first time the bankers, venture capitalists, and financial people had sat in on a meeting like this.

And it was the first time such a meeting had included space advocacy organizations such as the Space Transportation Association, the National Space Society, and the Citizen's Advisory Council on National Space Policy.

But not everyone in the room had experienced the paradigm shift.

A few aerospace company representatives not sitting at the table were heard to remark about the space advocates, "Where did these people come from? And why don't they go back there?"

In fact, the aerospace company people didn't seem to have caught on yet. They presented dissenting views to the concepts that Goldin and others were putting forth in which companies would design, build, finance, and provide spaceships to operating companies who would then, with private investment, fly these spaceships for anyone who paid for the service, including the government. In brief, the positions of *all* of the officers of the major aerospace companies present can be summarized and paraphrased in a simple statement: "Give our company large amounts of government money and a free hand, and we'll get the job done!" This general attitude did *not* go unnoticed by NASA Administrator Goldin, the people from Capitol Hill, and members of the investment community seated "at the end of the table."

However, everyone seemed to agree regarding the basic government role in new space ventures: Government involvement in the development of new, reusable spaceships is proper and necessary but requires a major shift in roles so that the government proves the technology and removes the perceived technical risks in order to induce private sector financial involvement.

A visionary statement came from Tom Rogers, President of the Space Transportation Association: "There can be little doubt that there are hundreds of thousands of people who would pay to go into space themselves if the next generation of space vehicles can reduce unit costs by an order of magnitude and provide reasonable safety and reliability."

Two years before, it would have been impossible to hold such a round table discussing the subject because only a few people at the table believed then that reusable, profitable, safe, reliable, aircraft-like spaceships were possible.

Max Hunter's "riot" was well under way and the work of the Citizen's Advisory Council was paying off.

NINETEEN

The CAN, the NAC,
and Other Players

ON OCTOBER 19, 1994, a draft Cooperative Agreement Notice was issued by the Program Development Directorate of NASA's Marshall Space Flight Center in Huntsville, Alabama. A Cooperative Agreement Notice is different from a Request for Proposal in which NASA says, in effect, "We want someone to develop or deliver a hypersonic thrimaleen, and we'll accept proposals or bids on such-and-such date, then select a winner." In a Cooperative Agreement Notice, NASA says, "We are interested in developing a thrimaleen in a joint effort with an outside industrial firm. NASA will put up X% of the money and the winning firm will be expected to invest Y% of its own money, keeping the proprietary technology themselves and thereafter being free to build and sell thrimaleens to NASA or anyone else who wants to buy them." The concept of the Cooperative Agreement is to let the federal government reduce the perceived technical risk so that the private sector will have enough faith to put money into using the technology, too.

The Cooperative Agreement was invented by NASA as an experimental way to fund the development of reusable launch vehicles. It was clear that both Congress and Dan Goldin didn't want Shuttle II. Both were convinced that the reusable SSTO was the way to go. They believed that the Cooperative Agreement approach would bring about true commercial space transportation. It may or may not work. After all, it's just as experimental as an X-vehicle.

The Cooperative Agreement Notice showed that NASA people were trying to say what their Administrator was saying while at the same time carrying on "business as usual."

However, the wording of the draft revealed that these NASA people hadn't deviated at all from their initial agenda announced after the first successful DC-X flights 13 months before. In addition, the approach they proposed wasn't very much different from that advocated in the NASA Access to Space report, also prepared at NASA Marshall Space Flight Center with the help of others at NASA Langley Research Center. The philosophy was basically unchanged from the Apollo and early space shuttle days.

In brief, the content of the document seemed to indicate that the people at these NASA centers hadn't caught on to the fact that their boss wanted a revolution and that, indeed, the revolution was going on. They didn't seem to want a revolution, much less even a restructuring, reformation, or minor revision of the status quo. *They* knew everything there was to know about space launch vehicles, and no one should argue with them because *they'd* learned how to do it from the German rocket engineers.

Cooperative Agreement Notice 8-1 covered the development of the X-33 Advanced Technology Demonstrator. When people in the SSTO community saw it, they immediately recognized the X-33 as the reincarnation of the X-2000 SSTO proposed by NASA in September 1993. In draft form, CAN 8-1 appeared to ask for Shuttle II.

The X-33 program was divided into three phases. The first was a study phase lasting 15 months during which two to three bidders would conduct paper studies of the X-33 they wished to build. Then NASA would make the decision before the end of 1996 whether or not to ask one of the contractors to design, build, and fly one of the proposed experimental vehicles. This would in turn lead to a government decision in 1999 as to whether or not to develop the next generation reusable launch system. To those of us on the Council, this sounded like Shuttle II in the making.

The Cooperative Agreement Notice also required the contractor to develop and submit business plans, financial plans, operating economic models, sensitivity analyses, market analyses, and other business and financial material that had absolutely nothing whatsoever to do with building an experimental follow-on to the DC-X.

Because the draft requested inputs from outsiders, NASA did indeed get them, especially from members of the Citizen's Advisory Council on National Space Policy who didn't have to worry about preserving their jobs or government contracts.

The final version of the X-33 Cooperative Agreement Notice, even after all the outsider comments were written into it, could still result in the X-33 becoming the prototype of a shuttle-sized spaceship that would be Shuttle II. Although the Notice specified no payload and didn't require that the X-33 achieve orbit, the vehicle asked for was *not* an experimental vehicle in the legacy of the X-1 and the X-15. The language could be interpreted to mean that the X-33 was the prototype of a space shuttle replacement. A prototype is *not* a true experimental vehicle. An X-vehicle is designed and built for no other purpose than to advance the state of the art, to build and break gadgetry, and to learn how to do things before designing a commercial product. The X-1 was not the prototype for a supersonic fighter plane or the commercial Concorde supersonic airliner. The X-24 lifting body was not the prototype for the space shuttle Orbiter.

Looking at CAN 8-1, many SSTO supporters had another embarrassing question that no one answered, "Didn't the contractors go through a study phase like this before in Phase I of the SSTO project back in 1991?"

Three companies responded to CAN 8-1: Rockwell International, Lockheed-Martin, and a new team, McDonnell Douglas with Boeing. When early details of the three X-33 proposals became known, it certainly looked like the three study phase contract winners had merely taken their 1991 proposals out of the archives and resubmitted them.

The McDonnell Douglas/Boeing X-33 entry was an evolved DC-X. One version that appeared in a set of McDonnell Douglas overhead transparencies had stubby wings because Boeing wanted a winged horizontal lander for a reason that will become clear in a moment. In September 1995, Livingston L. Holder, Jr., Deputy MDC/Boeing Program Manager said that his team didn't know what

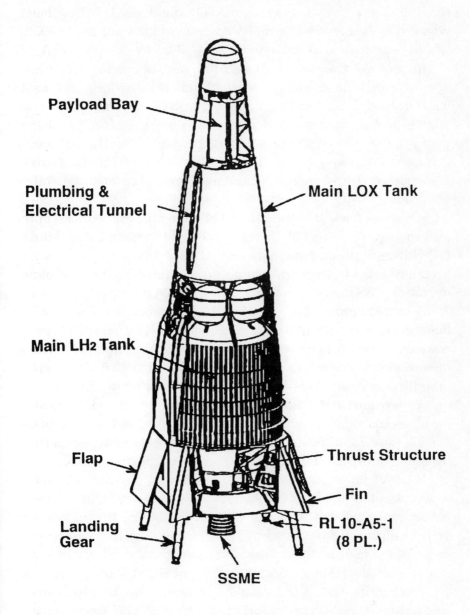

Payload Bay

Plumbing &
Electrical Tunnel

Main LOX Tank

Main LH2 Tank

Flap

Thrust Structure

Fin

Landing
Gear

RL10-A5-1
(8 PL.)

SSME

FIGURE 19-1: *A cutaway drawing of the McDonnell Douglas/Boeing X-33 Vertical-Takeoff-Vertical-Landing Reusable Launch Vehicle, "son of DC-X." (Drawing courtesy McDonnell Douglas Corp.)*

their final operational reusable SSTO would be. If MDC/Boeing won the X-33 competition, their experience with their X-33 vehicle would eventually determine that.

In contrast, the Rockwell X-33 proposal was a 50% scale version of a huge shuttlelike propellant tank with wings, a Vertical-TakeOff-Horizontal-Landing unmanned SSTO capable of carrying 40,000 pounds to orbit. Basically, it was what Rockwell had proposed in the 1991–1992 Phase I SSTO program. Max Hunter looked carefully at the weights and decided that Rockwell intended to use that elusive light-weight, high-strength aerospace material, "non-obtainium."

Although Lockheed had refused to bid on the Phase I SSTO project, their X-33 was a 60% scale version of the same Vertical-Take-Off-Horizontal-Landing "aeroballistic" spaceship that they'd privately briefed earlier to people in Washington. The full-sized operational SSTO version would carry 40,000 pounds to orbit.

All three proposed X-33 designs had "payload bays" that subsequently became "instrument bays" when NASA said that X-vehicles weren't supposed to carry payloads. Some of us wondered why the operational SSTO proposals showed a payload of 40,000 pounds to orbit. The answer: to carry up the space station modules. However, only a few space station payloads will weigh that much. So why build such a big spaceship? NASA's reply: "If the first SSTOs are feasible and more economical with a 10,000-pound payload, for example, the space station modules can probably be redesigned."

None of these proposals had been "frozen" by mid-1995. The reason was that the X-33 contractors were trying to figure out what NASA really wanted, whereas Goldin was telling them, "Give us your best shot, whatever it is." This contractor attitude was partly due to some insurrection in the NASA ranks.

The team at NASA Marshall Space Flight Center that had worked on the Access to Space study prepared a highly detailed five-volume engineering report on a winged SSTO using vertical takeoff and horizontal landing, a configuration they favored. They were told by NASA Headquarters to sit on it lest it influence the work of the three X-33 contractors. Instead, this group, under Uwe Heuter, quietly sent the report around to the three X-33

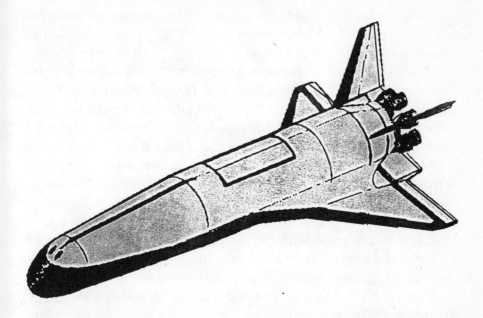

FIGURE 19-2: *Rockwell came into the X-33 Phase I competition with a huge winged tank operating in Vertical-Takeoff-Horizontal-Landing mode that would lift 40,000 pounds to orbit unmanned. (Drawing courtesy Rockwell International.)*

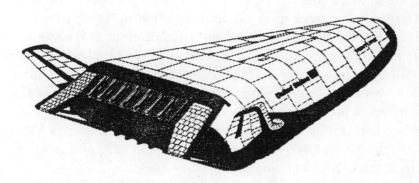

FIGURE 19-3: *The Lockheed entry in the X-33 Phase I competition was the aeroballistic Vertical-Takeoff-Horizontal-Landing ship they refused to propose in the original Ballistic Missile Defense Organization's SSTO study. (Drawing courtesy Lockheed-Martin.)*

contractors "for information purposes." The contractors took this as a signal that NASA really wanted an SSTO with wings like the space shuttle.

When several of us had breakfast with Goldin in April 1995, this was brought to his attention. The NASA Administrator was furious. A few weeks later, a definite signal was sent to the X-33 contractors: No one on that study team would be allowed to serve on the source selection board to determine the winner of the Phase I X-33 competition. Nothing else happened afterward to the members of this team who broke step with the Administrator. I met some of them who were still working at Marshall Space Flight Center in September 1995. It's nearly impossible to fire a civil servant.

It was taking some time and effort to educate NASA on how to do X-programs again. Basically, these programs involve building a series of X-vehicles that are *savable*, *fly often*, and *fly soon*.

It was also taking some time and effort to educate NASA people about how the commercial world works. Since its creation in 1958, NASA has operated as part of the Washington "command economy" that's different from the market economy of the rest of America. However, Goldin was trying hard to break the old NASA paradigms and therefore brought in many people from outside NASA to manage the Reusable Launch Vehicle program.

One of these was Air Force Colonel Gary E. Payton who was a military mission specialist on Shuttle Mission 51-C in January 1985. At the Ballistic Missile Defense Organization, he'd succeeded Sponable. When Goldin asked BMDO director Lt. Gen. Malcolm O'Neill to release Payton to NASA in early 1995, O'Neill declined to do so because it would cause him to lose a military officer's slot. So Payton, who saw no SSTO activity for him to manage there, resigned his Air Force commission and went to work for Jack Mansfield in the NASA Code X office, running the DC-XA, and X-33 programs from NASA Headquarters.

Goldin had become a "true believer" in the SSTO. He listed it as one of five major NASA programs: Keep the space shuttle flying, build the space station, encourage the development of SSTO as a cheaper and more reliable means of space access, explore the planets with cheap robotic spacecraft, and develop advanced aeronautical technology. Goldin is both loved and hated within NASA. But, in my opinion, he is the best NASA Administrator in a long time. Certainly, he's *trying* not to proceed on the basis of "business as usual." Whether or not he'll survive and be able to change NASA remains to be seen. I hope he succeeds.

In the meantime, the SSTO funding was pried loose from the Advanced Research Projects Agency, allowing the DC-X to be repaired and trucked back to White Sands in mid-1995. There it made four more successful flights in May, June, and July after which it was officially turned over to NASA.

However, the SSTO concept continued to draw sniper fire from opponents. NASA has an Advisory Commission composed of people from other government agencies, academic institutions, the science community, and the aerospace industry. Basically, Goldin was forced by political necessity to listen to what the NASA Advisory Commission (NAC) said. Members such as Dr. Jack Kerrebrock of the Massachusetts Institute of Technology were vehement opponents of SSTO. Several NAC reports severely criticized the SSTO program because NAC members didn't believe the technology existed to build an SSTO, didn't believe that an SSTO could be built that would carry a payload to orbit, demanded that "further studies" be funded before any decisions were made to build even the X-33, requested that the meager X-33 funds be reprogrammed to include two-stage-to-orbit (TSTO) studies and tests, and didn't foresee that the marketplace would support an SSTO.

As a result, Goldin formed the NASA Technology and Commercialization Advisory Committee under Jack Mansfield and got the NAC to transfer its oversight of the SSTO program to the new group. I was asked to serve on this new Committee, which I did from mid-July to early November 1995. In that period, I didn't learn anything I didn't already know about the SSTO program but I was able to present to NASA some of the information and material I'd gathered for this book. It was instrumental in silencing some of the NAC opposition to SSTO. But my Committee tenure only confirmed my fears that NASA might not succeed with the X-33 because Goldin was saying one thing and NASA people were doing something else.

In the meantime, 15 commercial telecommunications companies announced requirements to launch at least 1,385 satellites before 2005. This market greatly exceeded the existing space launch capability of the American, European, Russian, and Chinese space launch industries.

This has not gone unnoticed by numerous private firms such as Space Access, Inc. and Kelly Space and Technology who say they will attempt to get financing to build their private space launch vehicles.

Concepts such as Black Horse and Black Colt have been put forth by people such as Mitchell Burnside Clapp, a USAF test pilot now at Phillips Laboratories helping Sponable run the DC-XA flight tests. Black Horse was proposed as a manned aircraft powered by a combination of jet and rocket engines. It would take off from a runway with only jet fuel aboard, refuel in the air with rocket propellants from a tanker aircraft, then accelerate to orbit and eject its payload. The pilot then flies it back to the ground and lands at an airport.

Other companies were out there behaving in the typical American entrepreneurial manner. Many of them will not be heard from until they start putting cargos in orbit because they know only too well the truth of the business saying, "Macy's doesn't tell Gimbels." 90% of them will fail, as is typical of start-up companies. But practically all shy away from government funding and support. Many of them were started by aerospace managers and engineers who were laid off in the wake of the post–Cold War meltdown of the defense industry. They learned to talk to the financiers and investors. If those new companies wanted anything at all from the government, it was for the government to get out of the way.

This no-government approach was greeted with derision by most of the established aerospace companies. However, the existing aerospace firms may *not* be the ones who build the commercial spaceships. The makers of steam railroad locomotives didn't survive the introduction of the Diesel locomotive. In spite of Henry Ford's early activity in aviation that produced the Stout trimotor transport plane, the auto industry didn't spawn the aviation industry. It may well be that the existing aerospace industries may not be the major firms of the new commercial space age.

Bob Citron, then president of one of these new commercial spaceship companies, Kistler Aerospace Corporation of Kirkland, Washington, has openly stated, "We believe we can do it with all private financing when other companies in the aerospace industry require government financial support, government market guarantees, and government termination liability guarantees."

Citron is one of the founders of the highly successful SPACEHAB, Inc., maker of a manned laboratory module that can literally be plugged in to occupy one-third of a space shuttle Orbiter payload bay. SPACEHAB, Inc. is a profitable multimillion-dollar company. Kistler Aerospace Corporation was formed by some of the SPACEHAB investors after they'd seen the first DC-X flight. Its co-founder, Walter Kistler, is a highly successful entrepreneur in the electronic instrument business. However, in 1995, Citron and Kistler may have made a major error. They brought aboard Dr. George Mueller, the former NASA Associate Administrator for Manned Space Flight during the Apollo program. Mueller brought in a host of old NASA cronies such as Aaron Cohen, former Director of NASA Johnson Space Center in Houston, Texas. All of the younger engineers with the new entrepreneurial approach were fired. Gary Hudson, who'd developed a reliable rocket motor for the proposed Kistler K-0 vehicle, the company's version of the DC-X, had his contract terminated when Kistler changed the vehicle design to a two stage system. To those of us who remembered what happened to Hudson, Space Services, and the Percheron rocket, it was *deja vu* all over again. The Kistler K-1 rocketship is now envisioned as a two-staged fully-reusable vehicle capable of putting 2,000 pounds into orbit. Mueller says it will fly before the end of the decade. I hope it does. I hope my fears that Kistler Aerospace will go the way of Space Services prove to be unfounded. Commercial space transportation needs a successful private launch vehicle company that hasn't grown up building rockets for the federal government.

The crisis of change in space transportation continues and will certainly last for the remainder of this decade. After a third of a century of a national space program, space advocates finally learned something important: Anyone who wants to go into space is going to have to make it happen themselves. As space visionary and author Robert A. Heinlein advised in 1950, it's going to cost "your other shirt, your eye teeth, and your wife's wedding ring."

In space transportation, the first flight of the DC-X on August 18, 1993, was the turning point, the moment where it became obvious that there was no going back to the old ways of doing things. The fight against those who want to preserve the old ways isn't over, but we're winning because the new commercial space age holds the promise of space access for everyone so we can do new and interesting things in space and make money at it.

TWENTY

The X-33
Decision

(Author's Note: This chapter was added as close to publication as possible to inform readers of the situation with the SSTO program as of July 10, 1996.)

WHEN NASA TOOK OVER THE DC-X, the agency was anxious to use the "Little Rocket That Could" to test some new technologies. Therefore, during late 1995 and early 1996 at the McDonnell Douglas Huntington Beach plant, the DC-X was rebuilt into the DC-XA, the "Delta Clipper Experimental Advanced." The relatively heavy original propellant tanks were replaced with an epoxy-graphite liquid hydrogen tank, an aluminum-lithium liquid oxygen tank built in Russia, composite propellant lines and valves, and a host of new on-board electronics. These advanced materials reduced the weight of the DC-X by about 20%. A new aeroshell having the same shape but with new access panels was installed. Except for NASA markings, the DC-XA looked the same as before when it was rolled out at Huntington Beach on Friday, March 15, 1996.

The DC-XA was intended, among other things, to prove the use of aluminum-lithium and composite propellant tanks for future SSTOs. Its tanks were the largest ever installed in a flying vehicle.

Back to White Sands went the DC-XA, making its third trip and possibly logging a record number of highway miles. (Certainly, this road mileage far exceeds the distances it flew in tests.) It's the first and last reusable rocket to be able to do this; the X-33 and follow-on SSTOs will be too large for overland transport and will have to be flown from the factory where they're built.

At Clipper Site with Pete Conrad as Flight Manager, the DC-XA made its first flight in private on May 18, 1996. The first *public* flight was scheduled for Friday, June 7, 1996, and "all the usual suspects" showed up, including NASA Administrator Goldin, Associate Administrator Jack Mansfield, and General Malcolm Ward and his staff from the USAF Space Command. This was the first time that the USAF had shown any public display of interest in SSTOs, a factor that many of us considered highly relevant in view of later happenings.

The stewardship of NASA's Marshall Space Flight Center (MSFC) in Huntsville, Alabama, now the NASA center in charge of RLV development, revealed differences between the BMDO/MDC testing philosophy and that used by NASA over the last several decades. Apparently, landing the DC-XA vertically on a simple concrete slab wasn't considered "high-tech," so a new landing pad with a steel grate was built to channel the engine exhausts away from the rocket's base. On the first flight, the rocket exhaust flames didn't behave according to MSFC calculations and the south maneuvering flap—which had partly deployed due to a glitch in the new software—caught fire on landing. The White Sands fire department put it out without further damage and Burt Rutan's Scaled Composites, Inc., had a new flap ready in about a week.

To prevent another near-disaster, the MSFC people decided to land the DC-XA on the white gypsum sands next to the landing pad as had happened in the emergency landing of DC-X Flight #5. They reportedly hired a Ph.D. consultant to tell them how to make the gypsum sand more firm. The answer was, "Add water and compact it." This was done.

The public DC-XA flight went beautifully about 15 minutes ahead of schedule. But not before Goldin officially renamed the rocket "Clipper Graham" in honor of Daniel O. Graham who had passed away because of cancer on December 31, 1995.

When Clipper Graham landed on the watered and compacted sand, we saw a much larger cloud of white than had been evident during the previous emergency desert landing. We were allowed down to the landing site within 30 minutes, and it was obvious that the jets from the four rocket engines had dug holes in the sand that were *much* larger than on DC-X Flight #5.

The third flight was scheduled for later that afternoon, and some of us waited in the White Sands Public Affairs Office while a conference about what to do took place at Clipper Site. Many of us wished we could have been a fly on the wall as a result of what some sources later reported.

When the 5000° exhausts from the rocket engines had touched the compacted sand, the added water had exploded into vapor, causing the huge holes we'd seen. The MSFC project leaders wanted to cancel the quick-turnaround flight for that afternoon, study the situation for a few weeks, and build a new landing pad. USAF and MDC people pointed out that a quick turnaround flight was crucial to the program and the DC-XA could easily be landed on a 40-foot square of concrete next to the grate. The decision came four hours later. Suddenly, we were on the White Sands buses going back to Clipper Site where we stood around for an hour, watching thunderstorms develop all around us. At 7:15 P.M., Pete Conrad scrubbed the flight because (a) they'd run out of range time, and (b) thunderstorms might move in before a flight could be made.

The third flight took place the next morning, 26 hours after the second one. Some anti-SSTO people used the delay to claim that a rocket couldn't be turned around and flown again the same day!

As of this writing, the fourth and fifth flights of Clipper Graham haven't taken place.

In the meantime, it appeared that all was quiet on the political front in comparison to the dirty tricks and covert opposition that are reported earlier herein. As things turned out, this probably wasn't true.

Faced with declining budgets and a personal crusade to streamline NASA, Goldin continued to reorganize the space agency. Apparently, he couldn't touch such centers as MSFC, Houston's Johnson Space Flight Center, or the Kennedy Space Flight Center at the Cape. So the axe fell on NASA Headquarters in Washington. By the end of June, the word was out on the Internet "NASA RIF Watch"—RIF being the dreaded acronym for "Reduction In Force."

OSAT's Code X advanced technology functions are to be divided among various NASA centers while the space access function, including the RLV program, is to be upgraded to a full Headquarters office, probably with X-33 chief Gary E. Payton reporting directly to Goldin. Everyone else in OSAT must find new positions at NASA centers by September 30, 1996, or be "riffed." The spectacle of twenty-year NASA bureaucracy veterans arrayed against a decisive, change-driven Administrator would indeed be a circus except where the confrontation impacts and retards the SSTO development program that Goldin (and probably others, but for different reasons) want so badly. It will not be easy to break the power of the NASA elite that really doesn't want access to space for everyone. The validity of this contention is reinforced elsewhere in this book.

As discussed in the previous chapter, the selection of the contractor to build the X-33 was scheduled to take place in June 1996. McDonnell Douglas bid alone—without teaming up with Boeing—because the two companies couldn't agree on two things: (a) horizontal dead-stick (no power) landing with wings (Boeing) or vertical DC-X style landing with rocket power (MDC), and (b) who was the boss. The Rockwell and Lockheed-Martin proposals were essentially the same as previously discussed with minor changes. The proposals—in the form of computer-readable CD-ROMs instead of tons of paper documents—were submitted to MSFC in mid-May 1996.

The selection committee was chaired by Gary Payton and met continuously at MSFC for about a month. It was so intense that the word came, "Some members brought their golf clubs but never took them out of the bags."

The evaluation was conducted strictly within the bounds of CAN 8-1 and in accordance with an accompanying document titled, "Decision Criteria for the Reusable Launch Vehicle Technology Program, Phases II and III." The latter was jointly developed from the CAN by the Office of Management and Budget (OMB), the Office of Science and Technology Policy (OSTP), and NASA staff. Trying to analyze and make sense out of these two documents is like trying to decipher the IRS tax code. Apparently, one X-33 contractor figured it out and submitted its proposal accordingly.

The final decision was based on rating both the technology portion of each proposal and the "economic" or "business plan" portion. The words "economics" and "business plan" do not mean the same things to NASA and the aerospace industry as they do to everyday business people. In May 1996, a copy of the basic "economic study" of RLVs prepared by MFSC arrived in a plain brown envelope containing a computer diskette with a spreadsheet written in Excel. I gave it to my colleague Paul C. Hans, a merchant banker and investor who's an expert in this area and has also done considerable work with me as evidenced in Part III of this book. The feedback can be summarized as follows: "Anyone who dared to put this before any potential investor or banker in the real business world would be thrown out immediately!" Basically, NASA's approach to a business plan and marketing analysis was sophomoric, incomplete, and tuned to a world quite different than everyday commercial business.

On Friday, June 28, 1996, a meeting was held in the Administrator's office at NASA Headquarters. Present were Goldin, the NASA General Counsel, and selection committee chairman Gary Payton. Apparently, Payton told Goldin the committee's decision. The presence of the NASA General Counsel, I'm told, was to witness that Goldin did not make the decision but merely heard and approved the selection committee's decision as reported by Payton.

For four days, only three people (other than the selection committee itself) knew who'd won.

The decision was announced by Vice President Al Gore, accompanied by Goldin, at the Jet Propulsion Laboratory,

Pasadena, California, on Tuesday, July 2, 1996: Thirty minutes before the public announcement, representatives from the three bidders were informed privately of the decision.

It appears that NASA did not want to take any chances that one of the losing bidders might challenge the contract award in the federal courts because this would tie up the SSTO program for five years or longer and leave the decision in the hands of federal judges.

The decision obviously had White House approval because Vice President Gore made the announcement and, with Goldin's help, pulled the cover off a model of the winning design.

Lockheed-Martin Skunk Works won.

This sent shock waves through the SSTO activist community that had spent eight years bringing the SSX Vertical-Takeoff-Vertical-Landing concept, which is deeply rooted in SSTO historical development, from an idea to flying hardware in the DC-XA. The Lockheed-Martin Skunk Works winner was a radical departure from this and, in some ways, a continuation of the existing space shuttle operational paradigm.

Or was it? In the two weeks following the "down-select," additional information became available on the Lockheed-Martin X-33 and its follow-up operational RLV, the VentureStar™.

The X-33 is a half-scale version of the VentureStar™ which is interesting because, as we've seen, an X-vehicle isn't supposed to be a prototype. The X-33 is a triangular-shaped vehicle, 67 feet long with a 68-foot wingspan. In the last year, it has grown vestigial wings at the rear end, undoubtedly to improve the aerodynamic control qualities of this sort of a "lifting body."

Lifting bodies have a long history of development and test within NASA, going back as far as 1951 when H. Julian Allen and Alfred Eggers—at what is now NASA Ames Research Center—developed the theories behind the blunt re-entry shapes that made ICBM warheads possible. In an effort to improve the maneuverability of these bodies, Allen and Eggers discovered that, by cutting a cone in half lengthwise, it was possible for the shape to generate lift at high speeds. Hundreds of different lifting body shapes were tested in the Ames wind tunnels, leading to the concept of the "lifting body."

Except for symmetrical shapes such as spheres, all objects moving through the air generate "drag"—resistance to the object's motion through the air that slows it down. Depending upon the shape, it can also generate "lift"—a force at right angles to the drag force. Sport parachutists learned that even a human being is a lifting body that can be steered during its fall before parachute deployment. Granted, the lift force isn't much and the ratio of the lift to the drag (the lift-to-drag or "L/D" ratio) isn't very large. Even the legendary barn door will indeed fly if it slices through the air edge-on. Tip the front edge up a little, and the air pressure on the bottom becomes greater than the air pressure on top . . . and it flies if means are available to stabilize it in that orientation.

FIGURE 20-1: *NASA's choice, the Lockheed Martin VentureStar™, is shown poised for take off in this computer rendition. (Artwork from Lockheed Martin Skunk Works.)*

Work on actual lifting bodies began in 1961 as a bootleg project at the NASA Flight Research Center at Edwards Air Force Base, California, when some engineers and pilots, with limited funds of $30,000, actually built a manned plywood lifting body, the M2-F1. The M2-F1 was towed aloft by the only hot rod ever developed by NASA: A stripped Pontiac convertible, installed with the largest engine available, a four-barrel carburetor, four-on-the-floor, and the capability of towing the M2-F1 behind at speeds up to 110 miles per hour in 30 seconds. They painted it bright yellow and lettered "NASA" on both sides to keep people from thinking it was some private toy developed with federal funds. Some of the Mohave Desert highways in California and Nevada were used to calibrate the speedometer when the highway patrols weren't looking.

The first flight of the M2-F1 in 1963 was followed by the Northrop M2-F2 and HL-10 manned lifting bodies costing about $1.2 million each and towed aloft for a free glide to landing. Various stability problems were encountered at low speeds. In the '70s, television viewers were treated to the most spectacular of all lifting body accidents every week; Bruce Peterson's crash on May 10, 1967, was used as the teaser to "The Six-Million-Dollar Man." Peterson got into trouble with uncontrolled roll and crashed on the dry lake bed at more than 250 miles per hour, flipping over six times. He survived with the loss of only one eye.

In 1966, the USAF became interested in lifting bodies and, with NASA, contracted with the Martin Company to build the SV-5P, which later became the X-24A and the modified X-24B.

The successful piloted flights of these lifting bodies paved the way for the final configuration and the unpowered landing technique of the space shuttle Orbiter. Because of the space shuttle program, the NASA/USAF lifting body program came to a halt in 1975.

The Lockheed-Martin X-33 apparently revives this neglected history and perhaps some of its technology because, although Lockheed calls its X-33 an "aeroballistic rocket," it's a lifting body. Weighing 273,000 pounds, it will take off vertically, powered by a two-unit 400,000-pound-thrust Rocketdyne linear aerospike engine that was originally tested and shelved in 1971. It will be steered by differential throttling of the aerospike engine and will make a dead-

FIGURE 20-2: *On July 12, 1966, the Northrop M2-F2 lifting body completes its first flight at Rogers Dry Lake, Edwards Air Force Base, California, piloted by NASA test pilot Milton Thompson. The photo shows the lifting body just before touch-down with the landing gear extended. (NASA photo 66-H-1093 from Stine archives.)*

FIGURE 20-3: *The United States Air Force became interested in lifting bodies as a result of NASA research; it contracted Martin Marietta (now Lockheed Martin) to build the SV-5P shown here in 1966. (Photo by the Martin Company from Stine archives.)*

stick horizontal landing, steered by various aerodynamic control surfaces. Thermal protection will use a new and advanced metallic material. Reentering belly-first, the X-33 and the VentureStar™ will spread the heat loads over the belly. Its low L/D ratio means minimal ability to maneuver left and right of its course during re-entry. The X-33 is expected to reach speeds as high as Mach 15.

It is also an experimental vehicle with a payload bay—not an "instrument bay" as it was a year before—5 feet by 10 feet in size, but with no payload capability announced.

Flight testing is scheduled to begin from Edwards Air Force Base in March 1999. The X-33 will make about a dozen flights. Flight #1 is scheduled to go sixty miles northeast to Bicycle Lake Army Air Field near Fort Irwin. The second will fly northeast to Michaels Army Air Field at Dugway Proving Ground, Utah. The longest flights will go from Edwards to Malmstrom Air Force Base in Montana.

Apparently, the Lockheed X-33 isn't a self-controlled vehicle like the DC-X but an unmanned vehicle operated by constant radio link from ground stations along its flight path.

After each flight the X-33 will be returned to Edwards AFB by hoisting it with a crane atop the NASA 747 shuttle carrier aircraft and thus carrying it back just like the space shuttle Orbiter is ferried from Edwards AFB back to the Cape.

Lockheed-Martin confidently expects to build the VentureStar™. It will be 127 feet long with a 128-foot wingspan weighing 2,186,000 pounds at launch with a linear aerospike rocket engine producing 3,010,000 pounds of thrust. The payload weight has increased from 40,000 pounds a year ago to 59,000 pounds in the latest press kit—a whopping 47.5 percent increase.

Seeing these performance characteristics and numbers, plus the VTOHL operational mode, caused the author to revisit Chapter Thirteen of this book and look again at those of the X-2000 that NASA's Goldin and MFSC said they wanted in September 1993. It also caused him to look again at the numbers and operational characteristics of the space shuttle that is now wearing out. In addition, Lockheed-Martin continues to talk about its VentureStar™ as a replacement for the space shuttle.

Some Council members wondered how Lockheed-Martin will

be able to build this vehicle with such a complex shape and maintain the initial 0.901 propellant fraction. The structure is complex, using what Lockheed calls "conformal tankage." The composite tanks are both the structural members and retain the vehicle's aerodynamic shape. As an example of the concept, blow up several long, skinny party balloons. Twist one in the middle and bend it into a V shape. Then twist another one and nestle it inside the V shape of the first balloon, taping the two together. Whereas the internal air pressure of the balloons maintains their shape, the conformal tanks of the X-33 and VentureStar™ must carry all stresses and flight loads on such a structure using bulkheads and other internal tank stiffeners. This is a formidable design task. No other aircraft or space craft has used this concept before.

To what extent does this technology derive from the highly secret "Aurora" spy plane reported in Nevada and over the North Sea in 1991? This is unknown because the Department of Defense continues to deny the existence of such an aircraft. Lockheed-Martin Skunk Works reportedly built this mystery plane. It obviously exists because it has been seen and its sonic booms have rattled Los Angeles and Las Vegas. It's also been spotted on radar taking off from a base at Machrihanish, Scotland, where the radar operator was told to forget he'd seen it. The "Aurora" probably isn't its name, and it's probably a spy plane capable of operating at Mach 6 to Mach 8 far above the reach of any antiaircraft missile. If its technology contributes anything to the X-33, it may be in the areas of structures and thermal protection.

A further technical question is being asked: Why would NASA opt for a proposal that increases technical risks when simpler solutions are available? The answer seems straightforward to anyone aware of NASA's historical tendency to favor maximum new technology in any project. Administrator Goldin clearly said, "Give us your best shot!" To the NASA troops, this has always meant the shot with the highest and glitziest new technology. The laboratory shelves are stuffed with neglected technology whose only drawback was its perceived simplicity. The way the Delta Clipper DC-X was built with so little money in such a short time with a high success record is ample proof of this.

FIGURE 20-4: *A computer-generated mock-up of the VentureStar™ in orbit, the Lockheed Martin entry for the X-33 contract is a sleek and high-tech vehicle that uses lifting-body technology and is to be powered by a linear aerospike rocket engine. (Artwork from Lockheed Martin.)*

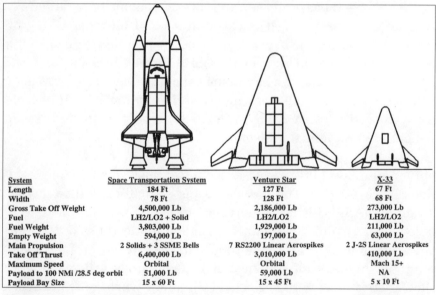

System	Space Transportation System	Venture Star	X-33
Length	184 Ft	127 Ft	67 Ft
Width	78 Ft	128 Ft	68 Ft
Gross Take Off Weight	4,500,000 Lb	2,186,000 Lb	273,000 Lb
Fuel	LH2/LO2 + Solid	LH2/LO2	LH2/LO2
Fuel Weight	3,803,000 Lb	1,929,000 Lb	211,000 Lb
Empty Weight	594,000 Lb	197,000 Lb	63,000 Lb
Main Propulsion	2 Solids + 3 SSME Bells	7 RS2200 Linear Aerospikes	2 J-2S Linear Aerospikes
Take Off Thrust	6,400,000 Lb	3,010,000 Lb	410,000 Lb
Maximum Speed	Orbital	Orbital	Mach 15+
Payload to 100 NMi /28.5 deg orbit	51,000 Lb	59,000 Lb	NA
Payload Bay Size	15 x 60 Ft	15 x 45 Ft	5 x 10 Ft

FIGURE 20-5: *A launch system comparison of the X-33 prototype, the proposed VentureStar™, and the current space shuttle shows how Lockheed Martin sees their SSTO as a "Son of Shuttle" program. (Artwork from Lockheed Martin Skunk Works.)*

That's the technical side of the equation. What about the "economic" or "business" side?

NASA intends to put more than $900 million into the X-33 program over the next three years. This was a given, regardless of who won.

The Lockheed-Martin proposal included a $220 million contribution that is a mixture of cash, in-kind use of existing company resources, and IR&D ("Internal R&D") money that includes general-purpose federal corporate technology base subsidies and contributions from various subcontractors. Apparently, the move that won the contract was a "pledge" by Lockheed-Martin to put up $2 billion dollars to build the VentureStar™ in the year 2000 and beyond. Apparently, Lockheed-Martin committed more of their corporate-controlled resources than either McDonnell Douglas or Rockwell.

Lockheed-Martin seems to believe it will be in a strong cash-rich condition by the end of this decade. It also plans to raise a bit more cash through short-term loans. After all, Lockheed-Martin already has a secure market position because it builds and operates the Titan and Atlas expendable launch vehicles with payload capabilities in the same range as the VentureStar™ as well as being a partner with Rockwell in the newly commercialized space-shuttle operations. This has already caused some marketing experts to wonder why Lockheed-Martin would build its own competition in a marketplace that will experience a space-lift shortage in the 1996–2006 time period. Others believe it was a move to head off the competition by attempting to monopolize the business. The real skeptics among the SSTO buffs have expressed the feeling that Lockheed-Martin can't succeed in building and flying the X-33, that they'll delay and stretch-out the program, and finally they'll announce that they've proved SSTO can't be done at all, thus putting away forever any challenge to the existing space transportation status quo. If Lockheed-Martin fails in the X-33 effort, it can recoup any losses with a few Titan, Atlas, or shuttle flights, admittedly a highly cynical view of the situation but certainly a businesslike one.

The Lockheed-Martin economic analysis includes a linear market study (see Chapter Twenty-Two) leading them to believe that a fleet of three VentureStars™ will more than handle the foreseeable space-launch market. These would be launched and recovered at the Cape every two weeks. The payload size was determined by NASA's requirements to launch about nineteen large space station modules. In April 1996 when I asked David M. Urie if Lockheed—who anticipates both building and operating the VentureStars™—would fly with smaller payloads, he said that was anticipated. But he didn't say if the full flight cost would be charged.

Some complaints have already been heard from the potential user community. The VTOHL mode of VentureStar™ means that satellites intended for flight on existing expendable launch vehicles would have to be redesigned. These satellites are stressed to take launch loads in only one direction (as they would also be in the VTOVL SSTO mode); putting them in a VTOHL vehicle means stressing for sideways landing loads in case of a vehicle abort or a failure to deploy the satellite in orbit.

Furthermore, even the huge 20-ton spy satellites of the National Reconnaissance Office are likely to be downsized to a larger number of cheaper 5-ton satellites. The 59,000-pound payload capability of VentureStar™ could go largely unused except for less than two dozen NASA space station loads.

Back in late 1995, Urie also told me that Lockheed-Martin would not build the VentureStar™ unless it received an iron-clad agreement from the federal government to launch all government payloads and obtain an anchor-tenancy agreement to build and supply the space station. This included liquidated damages for payload cancellations.

The Lockheed-Martin proposal reportedly included a deal to pay off its investment by offering NASA eight shuttle replacement flights over a two-year period at a price of $250 million to $300 million per flight, about two-thirds to a quarter of the cost of a shuttle flight (depending on whose numbers are used). It plans to fly twenty flights per year out of the Cape at a commercial price of $10 to $15 million per flight to cover fully amortized costs per flight of $4 to $6 million.

The Lockheed-Martin plan depends upon a 90% capture of the existing market. This is effectively a monopoly supported by

NASA, changing the players but not the basic business. This is likely to change over the next three to four years.

Many people who have been following the development of the new commercial space age feel that the shift to affordable, reliable space access would be better served by ongoing technical, corporate, and institutional competition toward lower per flight costs. The reasons for this are examined in Part III.

At first it may appear that the Lockheed X-33 proposal actually changes the incremental SSTO approach that we have been using for years: "build a little, test a little, fly a little." In other words, it appears we have gone to "build a lot, test a lot and fly a little." After the Lockheed X-33 selection, many of the vertical lander proponents felt it would be a waste to throw away six years of accumulated knowledge and testing. Early considerations have been given to taking the DC-XA and extending its life and test operations. This would not cost a great deal of money. For example, DC-XA testing could be continued through 1996 at an estimated cost of $3 million to the Air Force and $3 million to NASA. In the meantime, the DC-XB would be built using new tankage, a stretched aeroshell (make it a little bigger), a thermal protection system, and a fifth center engine. This could fly to Mach 3 in the summer of 1998 at a cost of $70 million for the Air Force and $10 million for NASA. The ultimate "stretch" would be the DC-XC with new conformal tanks; in other words, the aeroshell itself would be the tank walls. It would have an improved thermal protection system, lighter structures, and upgraded engines. It would fly to ten times the speed of sound in 1999 at a cost of about $130 million for the Air Force with the NASA cost dependent upon the desired NASA advanced technology tests.

Something like this might be a good thing for NASA, for the Air Force, and for the country. Here's why. The DC-XB/C complements the X-33 very well in terms of technology and institutional approach, exploring many known and promising RLV technical alternatives that are outside of the scope of the X-33. DC-XB/C would be an affordable hedge to the high stakes X-33 bet. While the X-33 will pioneer the use of complex multi-lobe composite propellant tanks, DC-XB/C could provide insurance against manufacturing and durability problems with much simpler geometry tankage.

Another factor is that X-33 would be oriented toward fixed operating bases with specialized ground handling equipment whereas the DC-XB/C would be aimed at more mobile operations out of small, austere sites.

Keep in mind that the Lockheed-Martin X-33 is a relatively high risk/high payoff approach, bundling a number of new technologies into a complex and sophisticated package. If it works, it's a great ship. But there is a lot of potential for delay and a lot of things have to come together all at once at the end of a very tight schedule. The DC-XB/C takes a more incremental approach. It may be that this dual approach is more politically doable, as well. NASA's top leadership endorses competing X-vehicles but has a bad budget to deal with. Whether or not we can afford to do the dual program approach is problematical. This is an era of tight budgets for NASA, even though the X-33 has major political support from the White House. A new DC-X vertical lander program in competition with the Lockheed X-33 horizontal lander program might also provide some interesting new competition. Furthermore, we wouldn't put all our eggs in one basket.

Regardless of what happens, three years is a long time. Things could change a great deal in that period. Given the promise of commercial space as detailed in Part III, it would not be prudent from a business standpoint to lock ourselves into one particular way of doing things at this particular point. Any good businessman always has plan B.

One thing is for certain: the next three to five years are going to be interesting.

PART III

ACCESS TO DREAMS

ONE THING SHOULD BE CLEAR at this point: *SSTO isn't going to go away.*

Pandora's Box has been opened. The idea is out of the box and loose in the world. Furthermore, SSTO is different from other space initiatives of the past because:

The reusable SSTO is a money maker!

Do not dismiss the economic motivation. Private space launch vehicle companies that understand the moneymaking nature of the new spaceships will build them without government assistance, regardless of what NASA does. Not all of them will be successful. The free enterprise capitalistic system only guarantees the right to try; it does not guarantee success.

But someone *will* be successful if they try. No one will be successful if they don't.

If private space transportation companies fail, this might mean another decade of business as usual within a NASA bureaucracy that endured in spite of an Administrator who wants to change it. However, regardless of who the Administrator happens to be at any given time, business can never again be "as usual" for NASA because Congress shows no interest in funding large space projects.

If for some reason American entrepreneurs don't achieve success in true commercial space transportation, it may come from another country whose businessmen understand what is discussed in Part III.

Don't count on Americans *not* doing it. An anonymous businessman once remarked, "Never underestimate the American response to a business challenge."

America opened this century by giving everyone access to the air. America can end the century and the millennium by giving everyone access to the solar system. Affordable access to space can and will be achieved.

But it will not be given to people by governments or eccentric billionaires who want to play with their money. It will be earned by people who work for it.

Space advocates did not get the futuristic world of the Kubrick-Clarke motion picture, *2001: A Space Odyssey*, because they thought the government was going to do it for them. The government didn't and won't.

Frontiers are opened by people, not governments.

They are opened because people seek wealth or the betterment of their lives and those of their children.

Frontiersmanship has always demanded taking risk with your life, your fortune, and your sacred honor.

But the rewards are worth it. The whole world becomes richer for it.

TWENTY-ONE

It's the Economics, Stupid!

IN DISCUSSING THE FUTURE of SSTO and its impact on space transportation, we assume that we'll be able to build reusable SSTO spaceships. The assumption is warranted on the basis of the long history of the SSTO concept and knowledge of the existing technology. Several experienced aerospace engineers with a known ability to "bend tin" and make things fly believe SSTO spaceships are feasible. Several independent companies believe they can build an reusable SSTO capable of carrying a payload to orbit and return. Four major aerospace companies—McDonnell Douglas, Lockheed-Martin, Rockwell International, and Boeing—do, too.

The big question is:

What will it cost?

Then the next question is:

Given the numbers, who will buy it and what can they do with it to justify the investment?

Some people claim these questions were answered satisfactorily for the space shuttle. They were not. In fact, it's now known that NASA people lied to Congress because they knew from the very beginning that the space shuttle would cost far more than announced and that it wouldn't be a moneymaker. That's history and it can't be changed. But keep it in mind when some anti-SSTO person dismisses the economic analyses of SSTOs, saying the same thing was said for the space shuttle. What follows is *not* the same sort of analysis.

Bankers, investment brokers, merchant bankers, financiers, and venture capitalists are the ones who must be approached for funding, not Congress. These people speak a different language, have different requirements, and represent a group that is alien to nearly all aerospace engineers and space advocates.

First, the financiers want an answer to the question: *How large a risk am I taking with my money, how quickly will I get my money back, and how rapidly will my investment appreciate?*

They'll ask whatever additional questions they deem necessary to get the answers, and they'll check these answers very carefully in a process known as "due diligence."

These people deal with real money, not dirksens obtained from the IRS.

They'll first evaluate the technical risks by calling on known experts to give them technical opinions. Sometimes, the expert will be Uncle George who used to work for ABC Aerospace Company. Usually, it will be someone in NASA. In the past, this meant the end of the road for budding spaceship builders because the answer was, "It can't be done." That answer cannot be defended today.

However, financiers may not want to invest while the government is engaged in building and flying SSTO X-vehicles designed to reduce the perceived technical risks. High technology risk reduction has been the historic role played by the federal government in aviation, for example. Fledgling aircraft companies built airliners using government research findings from the National Advisory Committee on Aeronautics (NACA, the predecessor to NASA). The DC-X has done much to lower the perceived technical risks. If NASA does what Administrator Goldin says NASA will do, *X-33 can do the rest.*

Some bankers may not believe that the basic technology is only waiting to be applied. Convincing them is a selling job, and prospective spaceship builders had better be prepared to make a very solid case for the fact that it can be done.

Basically, the banker isn't in love with the technology. This is a critical point to keep in mind. When a prospective space transportation company executive goes to New York City to talk to CitiBank, Prudential Bache, Morgan Trust, or whatever financial institution can provide the money necessary to build spaceships or operate a spaceline, *the banker doesn't care what the technology is*.

A banker wants guarantees, and these can be many different things. For example, if McDonnell Douglas, Lockheed-Martin, or Rockwell International needs a line of credit to build spaceships, they have a track record of making similar technologies work in a profitable manner. Thus, the financier assumes that such a company knows what it's doing with technology he doesn't need to learn about. This is a guarantee in the mind of an investor.

A new player is going to have a more difficult time, depending on the track records of the people involved. The banker will want to know if the new company is led by executives with outstanding histories of corporate financial management.

The bankers will also carefully scrutinize those companies who want to buy and operate spaceships. It's easier if companies like Northwest Airlines or Federal Express go looking for capital to expand their operations. They have track records. However, Fred Smith at FedEx had no such track record when he started, just a convincing business plan as well as a very large piece of his own money he was willing to put into the venture.

Realize that the future of commercial space activities lies in the answer to a simple question: Is this going to make money?

It's the economics, stupid!

With this in mind, Paul C. Hans and I began developing an economic space transportation model from the point of view of the investment banker. We anticipated the questions he'd ask the organizers of a prototype spaceline who are raising capital to buy a fleet of SSTOs.

Is there a market, and what is it? Assume for now that a market exists that can utilize the productivity of the SSTO. Markets will be discussed later.

What will it cost per flight to make money with an SSTO? Maximum payload capability is important but . . . most airliners operate at far less than their maximum allowable gross weight because they fly "bulked out"—i.e., they're jammed full of, say, goose feathers destined for the New England Feather Bed Company. However, a load of dense tungsten carbide billets may occupy only a small part of the cargo hold but bring the airplane to maximum weight ("grossing-out" the aircraft while there's still room in it). The important number to an operator is: *What does it cost to fly the bird?* That number determines what he must charge for the service.

This is a complex function of the costs of the vehicle, fuel or propellant, overhead, amortization, depreciation, taxes, profit margin, and all the other factors that figure into a business plan. The technology is immaterial but the basic technical assumptions must be announced at the start. A customer doesn't care what kind of airplane FedEx uses to deliver his parcel.

Furthermore, the exact operational modes—VTOVL, VTOHL, HTOHL, nose-first entry, tail-first entry, etc.—impact only the type of spaceport facilities required, not their extent.

This overall approach is warranted on the basis of the history of aircraft development and operation.

Putting aside the precise technology used for an SSTO, assume that the spaceship has the following basic characteristics:

1. It uses liquid oxygen/liquid hydrogen with no auxiliary propellants. Other propellants can and may be used. But the SSTO will be designed to use one oxidizer and one fuel. Adding another type of propellant for, say, a tripropellant rocket engine, increases the complexity and therefore the costs to buy the spaceship and turn it around between flights.

2. The airframe, propulsion, avionics, and other on-board systems are completely reused. Two-stage aircraft were tried and abandoned—e.g., the Short Maia-Mercury pick-a-back composite transatlantic aircraft of 1938. And Rocket-Assisted Take-Off (RATO or JATO) units also passed into the history books. The cost *and the flight safety aspects* of any spaceship that has to jettison parts in flight will always give the totally reusable SSTO a strong cost-effective edge.

3. It can be rapidly turned around between flights. A transportation vehicle doesn't make money when it's idle but only when it's carrying payload—literally, a load that pays—from Point A to Point B. Although the early Boeing 707s and Douglas DC-8s were far more expensive to buy and operate than the Lockheed Constellations and Douglas DC-7s they replaced, the jets could fly the same routes in half the time or less and could be used 12 to 14 hours per day because of their greater simplicity that led to faster turnaround.

4. The ground crew consists of only 50 to 100 people per flight. The fewer people on the payroll to keep the aircraft/spaceships flying, the lower the overhead. According to the Air Transport Association, most airlines have 75 employees per airplane; most of these people sell tickets.

5. The vehicle operates like an aircraft from austere "flight simplex" facilities. Ground facilities are capital-intensive although they can be amortized and depreciated. However, they aren't used except when the airplanes or spaceships are on the ground. Mobile skyscrapers used to service the space shuttle and other expendable space vehicles are also unique to a given vehicle type, whereas an airliner jetway isn't. The British designed their postwar transatlantic airliner, the Bristol Brabizon, to use existing 4,000-foot runways; Boeing didn't care that their 707 needed 8,000-foot runways because it wasn't difficult for even a Third World country to pour concrete and extend existing runways.

Given these, the next questions are:

A. What is it and what will it do? For the purposes of this analysis, a first generation SSTO spaceship will provide the service of transporting up to ten tons of cargo and/or passengers into orbit every three days.

B. What won't it do? A first-generation SSTO spaceship will not be configured to accommodate large, bulky payloads. It's a pickup truck, not an 18-wheeler. Payload support services for cargo will be limited as they are with current cargo airline operators. Customers will have to supply their own specialized support such as temperature control, pressurization, electrical power, etc. In addition, an SSTO will not operate from a large number of spaceports during the early years of the service.

C. When, where and how easy is the service to obtain? Each SSTO will provide its services at regular intervals of at least every three days. A full payload bay will not be required for a flight. Service will be provided from several spaceports. Weather will be no more of a problem than it is for subsonic commercial air transportation.

D. How dependable and how safe is it? The reliability and safety of an SSTO will be comparable to subsonic commercial air transport aircraft. An SSTO will not require significant maintenance between flights to achieve this.

E. How much will an SSTO spaceship cost? The largest available subsonic commercial transport airplane, the Boeing 747-400, costs about $150 million dollars, depending upon customer options such as avionics, cabin decor, etc. The smaller Boeing 737-300 costs about $50 million dollars while the new Boeing 777 costs about $120 million. Boeing makes money selling these airplanes at these prices. Furthermore, airlines buy them and make money flying them. At the start of 1991 Boeing had delivered or taken orders for more than 440 of the 747-400 airliner. By early February 1991, Boeing had delivered 2,000 Boeing 737 aircraft. We therefore made a highly conservative estimate based on numbers obtained from McDonnell Douglas that an SSTO would cost between $500 million and $1.5 billion dollars each, due primarily to smaller spaceship production runs. But the small number of spaceships many people anticipate may be highly conservative, as we will see.

Now we're on the track of determining the price per flight and the required revenue per flight, but we must still have the following financial information:

CAPITAL EQUIPMENT COSTS: In addition to the cost of the spaceship itself, what are the costs for ground support equipment and facilities?

OPERATING COSTS: What are the costs of marketing, G&A, insurance, spaceship maintenance, crew and ground personnel, and fuel?

FINANCIAL COSTS: What is the cost of using someone else's money? What is the competitive return on equity and cost of debt?

CAPITAL STRUCTURE OF THE BUSINESS: What is the company's debt to equity ratio?

OTHER PERFORMANCE DATA: What is the fleet attrition, anticipated spaceship life, facilities life, and flight rates?

The economic and financial model takes into consideration the above factors. It was designed to allow variation in inputs in order to determine those factors that had the most significant effect upon the required revenue per flight.

We made one other important and conservative assumption in the model: At the end of five years, the business was totally liquidated and all the good will, spaceships, and facilities were completely written off. The model shows no value for residual costs, market position, or used equipment.

Some technical aspects of meeting customer requirements will drive the configuration of the SSTO spaceship. Furthermore, in a first-generation spaceship, the differences between a commercial vehicle and a military vehicle converge. The cargo section can be adapted for the specific flight, military or civil. Thus, if the basic design of a first-generation spaceship can satisfy both customers— as was done with the Boeing 707/KC-135—a manufacturer can spread his nonrecurring costs over a wider production base. This in turn will reduce the cost of the spaceship and make it even more economically viable in the commercial sector. Furthermore, satisfying both types of customers through convergence not only lowers development costs but greatly reduces the risk involved in developing the technology.

The first scenario used the following key assumptions:

- Spaceship purchase price = $500 million.
- Ground Support Equipment costs per ship = $20 million.
- Spaceship life = 500 flights.
- Fleet attrition rate = 1% per year.
- Flights per year per spaceship = 100.
- Propellant costs = $0.50 per pound.
- Inflation rate = 5%.
- Debt/equity structure = 33%/67%.
- Tax rate = 40%.
- Competitive return on equity = 16%.

The results are impressive:

The total required revenue for placing ten tons of cargo and/or passengers into orbit using a reusable SSTO spaceship with the *given characteristics is $1.6 million (1994 dollars).*

This represents a price about 1% to 3% of current launch services.

This is what an operator must charge because it includes profit margin along with all the other cost factors.

But are the results reasonable? What assumptions exert heavy influence on the competitive posture of a reusable spaceship relative to expendable and semi-expendable vehicles?

To answer these questions, various sensitivity analyses identified those factors that have the greatest effect upon the required revenue per flight—that is, they identified the economic "drivers."

Inputs of particular interest were:

- spaceship purchase price
- propellant cost
- spaceship life
- number of flights per year.

The price of a spaceship will vary according to (a) the number of units produced and over which development costs will be recovered, (b) the recurring cost of production, and (c) the associated production learning curve.

The cost of propellants will depend upon the overall demand. The basic prices of liquid oxygen and liquid hydrogen depend upon the costs of the energy required to produce them.

The flight rate and unit life are determined basically by spaceship design. Spaceships that have longer life and allow faster turnaround between flights will influence the revenue needed to provide an attractive return on investment.

The results of the sensitivity analysis are interesting:

- A 500% increase in the price of the propellants from $0.50 to $2.50 per pound caused the required revenue per flight to increase from $1.6 million to $2.2 million, a 38% increase.
- When the propellant price of $0.50 per pound was maintained, but the price of a spaceship was increased from $500 million to $1.5 billion, the required revenue rose from the $1.6 million to $4.2 million, a 163% increase.

Of the two inputs, the purchase price of a spaceship clearly is the significant economic driver.

In the case of a fuel price of $2.50 per pound and a spaceship cost of $1.5 billion, the resulting required revenue of $4.7 million per flight remains a fraction of the cost of service using expendable and semi-expendable launch vehicles.

Other sensitivity analyses showed great flexibility between spaceship cost, spaceship life and spaceship flight rates.

The results of these sensitivity analyses are shown in Figures 21-1, 21-2, and 21-3.

For those who might believe that FAA certification costs could seriously skew the results, these factors were taken into account. The certification costs can be very high but they simply increase the basic spaceship cost from, for example, $500 million to $700 million, or from $1.2 billion to $1.4 billion. This means that the prospective operators will have to do a little more work when talking to the bankers.

To put development costs in perspective, it cost Boeing $185

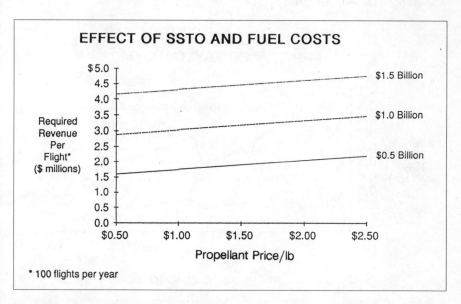

FIGURE 21-1: *Results of the sensitivity analysis of the effect of cost of an SSTO and fuel costs on required revenue per flight. (Courtesy Paul C. Hans and The Enterprise Institute, Inc.)*

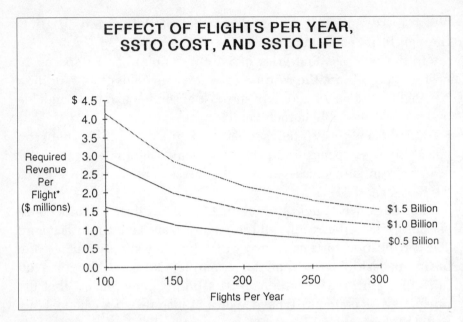

FIGURE 21-2: *Results of the sensitivity analysis of the effect of flights per year and SSTO cost on required revenue per flight. (Courtesy Paul C. Hans and The Enterprise Institute, Inc.)*

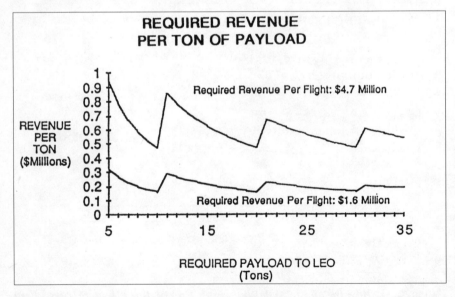

FIGURE 21-3: *Graph showing required revenue per ton as a function of required payload to orbit and required revenue per flight. (Courtesy Paul C. Hans and The Enterprise Institute, Inc.)*

million in 1954 dollars, $36 million more than the net worth of the company, to develop the Boeing 707; this amounts to $918 million in 1994 dollars. In 1970, the development costs of the Boeing 747 were $796 million of which CitiBank, the largest lender, put up only $34 million; the 747 development costs would be 2.745 billion 1994 dollars. Boeing recently stated that it put more than 6 billion 1995 dollars into the development of the Boeing 777.

However, a 1995 SSTO development cost estimate from a NASA contractor was $18 billion, which reflects the business-as-usual approach. Some high-level NASA and aerospace industry people believe that can be reduced to about $10 billion. Lockheed-Martin has quoted $6 billion for the development of their SSTO, and numbers from McDonnell Douglas placed the cost of developing Delta Clipper at between $4 billion and $6 billion.

Thus, we're not talking about a spaceship whose development costs are much more than a jet airliner. With government help in reducing perceived technical risk using the X-33, SSTO development cost could be about $6 billion, making the cost of an SSTO spaceship between $500 million and $800 million. If so, this makes the business opportunity even more attractive.

We ran the model with a spaceship cost of $2 billion just to see the results. In this scenario, the financiers are likely to conclude that the venture is slightly more risky. Therefore, they'll want the company to be more heavily equity oriented and with a greater after-tax return on equity of 20+%. In this case and making these changes in the financial structure of the company, the required revenue per flight became $10 million for ten tons. This is still extremely competitive against existing expendable launch vehicles.

The sensitivity analyses showed it's possible to have a great deal of variation in the first generation vehicles and still make money with them.

Also, the sensitivity analysis underscores what every airline in the world knows: *Keep the airplane flying.* The primary economic driver is the cost of the equipment and the efficiency of its use. Operating costs such as fuel become more important only when all competitors are using similar vehicles.

When payloads heavier or larger than those capable of being flown by a single spaceship are considered, the cost isn't a smooth curve. Figure 21-3 should be a familiar chart to people in the business of producing and operating civil transport aircraft. On the vertical axis is cost per ton where in the airline business it's normally cost per passenger. The spikes are caused by reaching the maximum payload capability of a single spaceship with a 10-ton capability where a customer wants to put up 11 tons. This creates a spike in the curve because an additional spaceship must be scheduled to lift either the additional ton or the payload divided equally between two spaceships.

But when the cost of orbiting large payloads in excess of the maximum payload of a single spaceship is compared to the same costs of lifting these payloads using the existing fleet of expendable and semi-expendable vehicles such as the space shuttle, Titan, Atlas-Centaur, Ariane, and Delta, the results are depicted in Figure 21-4. This dramatically shows that any reusable vehicle anywhere in the

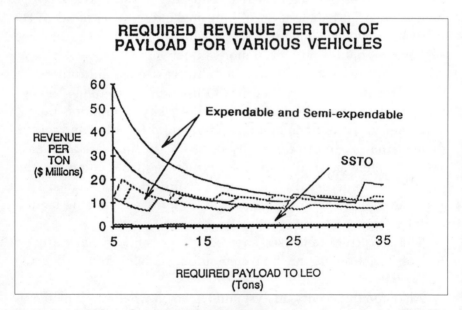

FIGURE 21-4: *When the required revenue per ton and required payload to orbit are compared against existing expendable and semi-expandable (the space shuttle) vehicles, the SSTO has such drastically lower costs that they can hardly be seen on this chart. (Courtesy Paul C. Hans and The Enterprise Institute, Inc.)*

neighborhood of the financial and operating characteristics of the SSTO spaceship used in the model is going to force a major change in the market.

Paul Hans and I continue to refine the economic model. And as SSTO spaceships come closer to operation and we get better numbers on costs, further analysis will define the exact nature of the costs, the necessary operational profile and the required revenue per flight.

However, the basic conclusion will still stand:

> *A spaceship with performance and cost characteristics in the ballpark with those outlined in this financial analysis will hold an overwhelming price advantage over any expendable or semi-expendable space transportation system.*

The flexibility of operation inherent in an SSTO spaceship will extend this competitive advantage still further.

The bottom line shown in Figure 21-5 is the situation producing a revenue of $1.6 million per flight yielding a total revenue over a five-year period of $915 million, a profit before income taxes of 25% of sales, and an after-tax cash flow of $720 million.

THE BOTTOM LINE

• – Revenue per flight	$1.6 million
• – Total revenue – 5 years	$915 million
• – PBIT as % of sales	25%
• – After-tax cash flow	$720 million

FIGURE 21-5: *The Bottom Line. Can a spaceline make money with a fleet of SSTOs? (Courtesy Paul C. Hans and The Enterprise Institute, Inc.)*

The financial analysis from an investor's point of view shows that a space transportation service using reusable SSTO spaceships can be brought to market at a price that can only be described as revolutionary.

*Lowering the price of a valuable service to between 1%
and 5% of current prices is not an event that goes unnoticed.*

And it has not. Hans and I were the first out of the gate with this economic model. Some people and companies have asked for the model; however, it's a proprietary piece of intellectual property and must carefully be tailored to a specific set of circumstances before investors can place a high level of confidence in the exact numbers. As mentioned above, the numbers are in the general ballpark. However, as the commercial space transportation industry gains momentum and as several companies enter the field, a few percentage points in the business planning can result in beating the competition or being beaten.

In spite of the fact that the technology allows us to do a reusable SSTO, the economics give an even greater sense of excitement. Technological capability alone isn't enough, as the space shuttle showed. But when the economics enter the picture in such a sanguine manner, it presents an even more convincing case that SSTO is indeed the vision that will create tax revenues instead of spending them and turn commercial space transportation into a win-win game.

People took notice when we presented these results before several professional meetings and in peer-reviewed journals. The early reactions to our economic model have spurred other studies that have confirmed our findings. One of these was the comprehensive Commercial Space Transportation Study (CSTS) issued in May 1994 by the Aerospace Alliance made up of Boeing, General Dynamics, Lockheed, Martin Marietta, McDonnell Douglas, and Rockwell.

It has also started people thinking about markets.

TWENTY–TWO

To Market,
To Market

NOT VERY LONG AGO, people didn't want space transportation. Most people today don't. They couldn't care less because they consider that space is a place for superhuman astronauts. They don't care that more than 60% of the world's telephone calls and all of their worldwide television coverage goes via satellites. They will say this while watching sports on ESPN or the latest news on CNN, both of which depend upon satellites.

People who want space transportation usually already have it because they need it to carry on their businesses although they will quickly admit they're paying too much for a risky ride to orbit. However, some users don't have it yet, and it's frustrating their ventures. Motorola and its consortium in the Iridium Project had to lock up in advance practically all the space launch capability in four nations for years to come. Ronald F. Taylor, a program manager at Motorola, Inc., and a fellow member of the Arizona Space Commission, told me that 30% of the cost of the Iridium project was set aside for launching the satellites, *not* for the satellites or the ground equipment that would bring cellular phone service, fax, and digital data communications to anyone anywhere in the world, including the North Pole, the top of Mount Everest, or deepest, darkest, dankest Africa. (*Distribution* magazine reports that transportation costs normally amount to between 2% and 5% of the final product cost.) If any of the launch providers in the United States, the Peoples' Republic of China, Russia, or Europe get into trouble with their

Delta, Long March, Proton, or Ariane launch vehicles, it means expensive delays for Motorola.

Bill Gates' Teledesic venture needs rides to orbit for 1,000-plus satellites. Hughes Aircraft Company requires space transportation for their Spaceway system.

As of 1995, telecommunications companies alone plan to launch 1,385 satellites into low-Earth orbit before 2005 A.D., according to such market studies as the Commercial Space Transportation Study.

Therefore, a market for transportation of unmanned satellites to orbit exists.

According to studies and data from NASA, the U.S. Department of Commerce, and The Enterprise Institute, Inc., the U.S. no longer dominates the space transportation market. In 1974, the United States enjoyed a 100% market share. In 1995, this had shrunk to 30%. In 1994, this market was estimated to be $460 billion. Conservative market estimates in the Commercial Space Transportation Study (CSTS) final report showed that this could grow to *$3 trillion per year by 2003* if only expendable launch vehicles were available and used.

The question arises, of course: Where is the launch capability going to come from if it's constrained by the number of launch pads on which expendable launch vehicles sit for as long as six months prior to firing?

When the reusable SSTO concept and its economics are first brought before most aerospace engineers, the reaction is, "There's no market for a system that will fly ten tons to orbit every day!" They see a flat market in spite of the data presented above and in the CSTS.

Some people don't understand marketing. Therefore, before going any further in the discussion of markets for commercial space transportation, I want to ensure that readers understand what I mean by the word "market."

"Market" comes from the Latin word *mercatus* meaning "trade," a traffic in the exchange of goods, services, or valuta between people. Trade means getting what you want in exchange for giving someone else something they want.

I learned three important marketing principles in my training by people who had formerly worked for General Electric Company, perhaps the corporation with the most effective marketing in the world. Later, I successfully used them as the marketing manager of an electronic instrument company.

The first says that marketing is *not* sales although sales may be part of marketing in some companies while in others marketing is a servant of the sales department.

The second states that there's a difference between a *want* and a *need*. People probably don't *need* what you want to sell or trade for something they've got. Even if they do need it, they may not realize it. Or they may recognize they have a need but want to buy a competitor's product instead. You have to convince them to buy what you've got.

The third principle maintains there are two ways to do marketing: (1) You can enter an existing market and figure out a way to sell what you've got by convincing people it does something for them that other products or services don't. Or (2) you can create a market where none has existed before by convincing people that your product or service does something for them that nothing else does or has ever done before.

In both cases, you must create a *want* or an urge to buy.

Keep these principles in mind with respect to the coming age of commercial space in which the introduction of reusable commercial SSTOs with their greatly reduced costs and improved services will trigger a change that can only be described as revolutionary.

Whenever such a major change is brought to a market, market dynamics change to a degree that causes major ripples through other markets. A good example of this is the introduction of communications satellites and their effect on the television industry and television programming as well as their impact upon long-distance and intercontinental telephone service. The rapid decrease in the cost of high capacity microchips and integrated circuits allowed the development of personal computers that created major changes in the computer industry and the manner in which companies, both large and small, conduct business.

Similarly, the introduction of SSTO spaceships into commercial service will revolutionize the infant commercial space industry. Regular and inexpensive transportation services to orbit will cause a major shift in the design and operating philosophy of satellites, space stations, space-based production facilities, Earth-lunar travel and even interplanetary travel.

For example, an SSTO that is refueled in orbit has the capability to fly to the Moon, land, lift off, and fly back without additional refuelling. Crewed flights to Mars can use the same SSTOs that, because of their vertical landing ability, can touch down on any celestial body in the solar system having a solid surface. The ability of an SSTO to change its velocity by 36,000 feet per second means that explorers don't have to "fly economy class" to Mars because the trip can be made in far less time. Robert Heinlein was right in 1950: *Get to low-Earth orbit, and you're halfway to anywhere in the solar system.*

But let's concentrate on working in our own backyard, Earth orbit, because that's where entrepreneurs can conduct the initial operations that will provide a near-term return on investment. Someone or something must pay for this space access capability, and the best way to do this is to allow individuals to use it to provide valuable products and services for other people. Some of them are already doing this, but at a higher price than will be possible within a decade.

Present satellite design and operations are costly. This is the result of three factors: (1) the high cost of getting to orbit, (2) the lack of a significant manned presence in space, and (3) the long and uncertain lead times in obtaining space transportation services.

Of the three factors, reducing the cost of reaching orbit has the greatest impact. If the cost is substantially reduced, the heavy emphasis placed upon the weight and redundancy of satellites, space stations, and other space facilities will be reduced, too. This decreased emphasis on the very high reliability required by the one-shot operating philosophies of expendable launch vehicles will permit the utilization of less exotic materials and the application of modular spacecraft design. These factors will lead to less expensive spacecraft and an increase in market demand for the services that can be provided by cheaper spacecraft and space facilities.

Low-cost access to space will also allow more energy-intensive flights for orbital, translunar, and interplanetary travel as mentioned above. The resultant savings in interplanetary flight transit times will decrease the need for spacecraft system redundancy and, in turn, lower the cost of these spacecraft and the trips.

A greater manned presence in space will also have an effect upon satellite design. The ability to reach a satellite easily and quickly will permit on-orbit repair and lower the need for heavily redundant fail-safe space assets. Like any commercial operation on Earth, a space-based commercial operation will provide the greatest return on investment when it's being used most effectively. The inability to reach an inoperative asset in orbit means the asset is "dead" from an investment standpoint. The ability to reach that asset quickly to repair it turns on the cash flow stream once again and lowers the burden of capital and fixed operating expenses. This, in turn, allows a lower cost to the ultimate beneficiary of the space-based operation and a greater return on investment to the space asset owner.

The specific shape and extent of the SSTO space transportation market curve is largely unknown. *The danger lies in underestimating the effect of the reusable SSTO spaceship.* Although the exact market dynamics that lie ahead cannot be determined right now, some estimates can be made concerning the extent that commercial spaceship operations will affect demand and the infrastructure needed to service this demand.

Using present space traffic models based on expandable launch vehicles really isn't valid.

As Stephen J. Hoeser pointed out in the Spring 1994 issue of *The Journal of Practical Applications in Space*, "In projecting future launch markets, there has been a tendency to perpetuate rather than speculate."

Brian Hughes, President of American Rocket Company, commented at a breakfast meeting of the Space Transportation Association in January 1994, "You don't size a bridge for trucks based on the number of people who are currently swimming the river."

But this is what most space market gurus are doing. Given the present cost of lifting cargo to orbit and looking at the "known" launch plans of various government agencies and commercial users based upon expendable and semi-expendable launch vehicles, the total demand for launch services by the end of the century has been estimated to be about 1.25 million pounds of payload per year. Applying a growth rate of 5% per year to this amount for ten years yields a launch requirement of about 2.0 million pounds of payload to orbit by the year 2010.

However, the introduction of the reusable SSTO spaceship will bring totally new economics to the space transportation business and open space to private enterprise in a major way.

Determining a market size for SSTO spaceship services is uncertain since the market demand curve is largely unknown, but we can make a WAG (Wild, Assumed Guess) at it, based on some new math.

It's almost a "field of dreams"—i.e., build it and they will come. During its growth period in 1988, America West Airlines did just that. They established round-trip airline service on a daily basis between Phoenix, Arizona, and Des Moines, Iowa. Other airlines thought America West was crazy because they "knew" there was no market there. America West replied, "If we put in the service, it will create a market." And it did.

Ron Taylor of Motorola, Inc., remarked during a meeting of the Arizona Space Commission in 1994 when I presented some of this marketing data, "Sometimes you decide to go ahead and do something because it feels like the right thing to do. Maybe the marketing studies support the decision, maybe not. Risk is always involved, but how you evaluate risk is always subjective and personal."

One can logically and reasonably assume on the basis of historical evidence that a sharp drop in space transportation costs will produce a proportional increase in the demand for space transportation services. An SSTO spaceship transportation cost that is 5% of services using expendable and semi-expendable vehicles indicates a potential but highly conservative annual requirement of 40 million pounds to orbit by 2010. It could be more. It could also be less if NASA studies the SSTO to death or if entrepreneurs cannot get their start-up capital. In the vernacular, we could screw up big time and leave the field open to others beyond the oceans who will read this and who can also do market and economic analyses.

Of course, many factors can affect this early estimate for SSTO spaceship launch services. As the aircraft and airline industries learned during the 1930s, estimating errors are more likely to result from conservatism when new technologies and major shifts of economics are introduced to a market.

A look at some modern air cargo figures can help put the estimated 40 million pound launch forecast into perspective.

Data from *Distribution* magazine showed that in 1988 the three major airports of the New York metropolitan area processed a combined total of more than 1.8 million tons of air cargo. That's more than 3.6 billion pounds of cargo for just one region of the world. Assuming that the air cargo market continues to grow at its steady historical rate of 2.5% per year, by 2010 the air cargo volume for the New York area would be about 6.2 billion pounds. The region can handle this increase because several New York area airports such as Stewart, Suffolk County, Bethlehem-Allentown, and Bridgeport Sikorsky are under-utilized.

Placed in this perspective, the estimated 40 million pound space launch requirement of 2010 represents less than 1% of the projected air cargo volume of only one region of the world.

With the rapid growth of space-based manufacturing operations expected to be triggered by the introduction of SSTO spaceships, the Earth-to-orbit cargo transportation requirements will grow rapidly. Considering the historical perspective provided by air cargo, it's not fantastic to speculate that the annual space transportation services required during the first two decades of the next century will grow to be far in excess of the 40 million pounds of payload initially estimated. How much in excess? This is truly anyone's guess, but the numbers could be in the hundreds of millions of pounds.

Obviously, this wide range of possible futures has a bearing on spaceship fleet size and the required ground infrastructure needed to support this magnitude of operation.

Assuming that each SSTO spaceship is capable of carrying 10 tons to low-Earth orbit and has a flight rate of 100 flights per year, the total fleet size required to meet an estimated market demand of 40 million pounds of payload to orbit is 20 spaceships. It's unreasonable to assume that a spaceship will be suitable for every payload. In addition, every spaceship won't have a full cargo bay on every flight because not even FedEx or UPS operates "grossed-out" or even "bulked-out" every night. For the purposes of this analysis, it's assumed the market requirements are such that spaceship suitability for cargo is offset by a cargo bay less than full. The result is a fleet size requirement that remains at 20 spaceships.

For a market requirement several times larger—such as 120 million pounds of payload to orbit—spaceship fleet size would increase to 60. Does this sound unbelievable considering the level of space transportation today? Not really. Orbital space promises to be an ideal environment for the manufacture of many high-technology products such as electronics and pharmaceuticals.

But some of the best products from an investment standpoint will undoubtedly turn out to be the mundane or appear to be "frivolous." As the cost of access to orbit decreases, the importance of the transportation portion of a space-based investment decision becomes less important.

A look at the manifests of the air cargo carriers in and out of the New York area shows that a large portion consists of such "low-tech" items as clothes, toys, games, vegetables, fruits, and seafood.

To say that space will forever remain the domain of high technology and highly-trained specialists is to ignore the business lessons of history and the economic promise of fully reusable spaceships in frequent and dependable service.

Some people may object to this analysis on the basis that much the same sort of thing was claimed for the NASA space shuttle. However, the major difference revolves around the fact that the designers and builders of the space shuttle didn't have to talk to New York bankers but to people in Congress who are quite a different class of financiers.

An example will clarify this. Suppose one goes to a New York investment banker and says, in essence, "I want to build and operate five airliners designed to a new concept. I'll fly them from Los Angeles to New York. When one of them gets to New York, I'll take off the wings and throw them into the bay. I'll remove the engines and ship them to Connecticut to be completely overhauled and ready to fly again in only three months. And I'll truck the fuselage back to Los Angeles where it will be rebuilt for the next flight six months later. If I can book thirty tons of cargo on each flight, I can make six flights per year. And I'll be economical; I won't fly one unless it carries a full payload." No investment banker will pay attention to an idea like that. If one looks at the space shuttle as an airliner, as I just did here and as NASA originally claimed they did, it becomes obvious that the costs are enormous and the chances for developing a going business flying people or cargo between Los Angeles and New York with this system are more than mildly risky.

This is because the driver in the case of any sort of capital equipment is to use it and reuse it as often and as much as possible since it doesn't make money while it's idle and accruing interest charges on its debt.

This is also the driver in the case of any spaceship. The secret of making money with any transportation device is to keep it in service making money. That is how airlines counteract the $150 million purchase cost for a Boeing 747-400, for example. For the space shuttle with a multi-month turnaround time and a throwaway of major parts on every flight, it simply isn't possible to fly it for less than ten or more times the projected cost of an SSTO spaceship.

Certainly, the market will react to the low-cost, reliable, regular, on-demand service provided by the SSTO. We don't know the extent to which it will react because we don't have the mathematics yet to reliably handle nonlinear systems, especially those that exhibit a revolutionary discontinuity when disturbed. We're beginning to understand some of this, thanks to the new principles of chaos theory.

An example of this is shown in Figure 22-1, which some readers will immediately recognize as the population curve of chaos theory. Biotechnologists know it as the growth curve of bacteria in a Petri dish. I've known it for years as the "Gompertz Curve." Hoeser showed that it applies to the nonlinearity of the SSTO orbital market—and practically any market, as was obvious when I read his paper in the Spring 1994 issue of *The Journal of Practical Applications in Space.*

As long as costs are high, demand stays low. When costs are lowered, demand increases. This is classical economics and marketing. It's difficult—if not impossible—to forecast what happens to demand when costs are suddenly and drastically reduced as shown. The situation becomes nonlinear.

This fact was brought forth in the May 1994 Commercial Space Transportation Study referenced earlier. Using the old terminology of "dollars-per-pound to orbit," this conservative study determined that when costs dropped to $600 per pound, the curve went nonlinear, making the classical market projections completely invalid.

Our study showed that a reusable SSTO could lower the cost to $80 per pound or less, using the old standard for comparison only.

Classical marketing know-how says that reliance on the "field of dreams" approach for SSTO services is going to be a crapshoot, as Motorola's Ron Taylor remarked. It will be viewed by some people as very risky. They may opt to wait and see. And they may lose out as a result.

Chaos theory now indicates that the way to get through such difficult nonlinear periods is to push very hard, making the slope of the curve as steep as possible. If this isn't done, the curve doesn't inflect at the top end and become linear again; it turns downward instead. Readers who have access to the computer software based on James Gleick's best-selling book, *Chaos: Making A New Science* (New York, Penguin Books, 1987, ISBN 0-14-009250-1), or the book itself can check this for themselves. I have, and it was an eye-opener. Until one has delved into chaos theory, it doesn't sound reasonable. However, my reaction was, "But of course! Why didn't I see it before? It's intuitive!"

Other people have seen the same thing in other areas of science, technology, and business. For example, in the words of Dr. William H. Calvin, the anthropologist-author of *The Ascent of Mind* (New York, Bantam Books, 1990, ISBN 0-553-35230-X):

> In evolutionary arguments, it is no longer enough to demonstrate that something should have done the job, given enough time. By compound-interest reasoning, *any* slight advantage can eventually do the job. There are usually

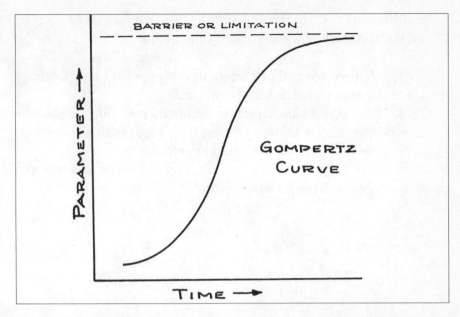

FIGURE 22-1: *The classical growth curve, often called the Gompertz Curve, fits the performance of system concepts as they evolve, the population of bacteria in a Petri dish or people in the world, and market growth. (Drawing by G. Harry Stine.)*

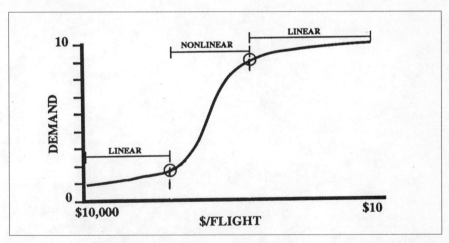

FIGURE 22-2: *The Gompertz Curve applied to market growth for SSTO services. When the cost per flight is high, the demand is low and slowly increases as costs are reduced. Then at about $600 per pound, the curve rises suddenly and gives birth to non-linear market growth. (Drawing by G. Harry Stine from Hoeser et al.)*

multiple ways to do the job, and the one that gets there first on the fast track tends to preempt the niche.

From what we see and can forecast, this appears to be precisely what's happening in space transportation.

By looking at just the economic numbers, it should be obvious that the people who build and operate SSTO spaceships will dominate space for the first half of the next century.

However, transportation of cargos to and from orbit won't be the most profitable or largest market at first.

TWENTY–THREE

Single-Stage-to-Overseas: Global Express

SEVERAL YEARS WILL PASS before the space launch markets adjust to the introduction of the new spaceships. Business moves slowly and cautiously with any new concept until it gets a track record.

Do other markets exist that could use the services of an SSTO spaceship or that *are created by the existence of such a spaceship*?

In short, if we can overcome market fixation, there may be markets other than the most obvious one just as there are many ways to build an SSTO spaceship. For example:

> *If an SSTO spaceship can carry a payload to orbit, it can carry a payload of similar size to any place on the Earth's surface in less than an hour.*

Figure 23-1 shows a polar projection of the world centered on the southwestern United States. The oddly-shaped circles with numbers in them depict the time in *minutes* it takes for an SSTO to deliver a payload to those regions.

A 15-hour flight to Tokyo becomes a flight of less than an hour if the vehicle can take off, climb above the atmosphere, cruise through space, and come down to land at the destination. Furthermore, it's environmentally benign. Instead of attaining the speed of 36,000 feet per second to fly into orbit, the SSTO can use less rocket propellant because it has to reach a speed of only about

FIGURE 23-1: *A polar projection of the world centered on a hypothetical Southwest Regional Spaceport in the United States. Circular areas containing numbers indicate the number of minutes of flight time required to fly there from the Southwest Regional Spaceport with Global Express. (Drawing from an SDIO/BMDO briefing paper.)*

24,000 feet per second to go halfway around the world and land. Thus, the SSTO becomes a Single-Stage-To-Overseas cargo rocket.

Expendable space launch vehicles can't do this safely or economically. Imagine your FedEx shipment being shot at Tokyo, plunging into the atmosphere over Hokkaido like a warhead, separating from the disintegrating totem pole that launched it, and descending to the ground in the vicinity of Tokyo under a parachute. Is there a market for such a service? It doesn't take a rocket scientist to answer that question.

The departure, climb-out, entry, landing conditions, and other flight operations of flying Earth-to-Earth are identical to that required for the true SSTO going to and from orbit. This Global Express system can use a derated SSTO spaceship rather than one specifically designed for the task. If SSTOs are going to cost two to three times that of a Boeing 747-400, an SSTO operator will want to use his fleet any way he can to keep it flying and making money.

And while the Earth-to-orbit market is going through its nonlinear evolutionary phase as discussed in the previous chapter, the SSTOs can be used for Global Express.

While waiting for your ship to come in, you might as well be digging for clams.

Global Express may be the profit center in the first five years of commercial space transportation while current and potential satellite builders/users become accustomed to the new facts of space transportation.

But will customers be willing to pay the price for Global Express service?

Again, it doesn't take a rocket scientist to show that people have *always* been willing to pay the additional price—often reluctantly until using the product or service becomes essential to the conduct of business or the maintenance of personal life-style—for anything that saves the precious and irreplaceable commodity of time.

Global Express is an application for SSTO spaceships that has been largely overlooked or ignored by SSTO advocates. *Successful* space transportation companies won't ignore it because it's going to pay the bills and work down the capital debt during the first few years while the space user community becomes accustomed to the new operational paradigm.

But where are the markets? Where are the users of this service?

We need to look at the *flight paths* over the surface of the Earth that will be followed by SSTO spaceships in Global Express mode. Understanding these paths can be extremely revealing to those who want to build, fly, and make money with commercial SSTOs.

These flight paths are great circle routes.

Many transoceanic (and even transcontinental) aircraft today fly great circle routes.

A great circle is the shortest distance between two points on Earth. Flying a great circle course was once a grand exercise in navigation because of constantly changing headings. Today, plotting and following a great circle course is done easily, thanks to the Global Positioning System (GPS) network of NavStar satellites and small, on-board computers. I have two commercial aviation flight planning computer programs that determine great circle courses and distances. These were used to create the maps and perform the distance calculations that accompany this chapter.

Satellites launched into orbit also follow great circle paths if one ignores the variations caused by Coriolis effect—the fact that the Earth turns underneath the satellite while it goes around the planet.

Therefore, great circle paths will be the ones flown by Global Express vehicles. Only a few degrees of heading, a few miles of ground track difference, and a few seconds of time are different if one takes Coriolis effect into account. So it will be ignored here in favor of making a point that's valid even when Coriolis effect is introduced into the equation.

An infinite number of great circle routes exist. Does this mean that SSTOs will fly everywhere?

No, not if the companies who own and operate them intend to make money.

Their customers are in the highly industrialized areas of the world. This is because "time is money" in the industrialized market economies.

Consider the following scenario:

At 3:07 A.M. on the graveyard shift of the Great New South Wales Paper Products Company Ltd. in Australia, the steam line feeding the drier rolls on the high-speed newsprint machine broke. Live steam fried the main control computer along with its data input converters. The paper machine stopped, creating a mess. Fixing the steam line and cleaning up the machine are messy and take a few hours. Getting a new computer isn't as easy. Spare boards and other components are kept on hand for ordinary emergencies, but the steam leak cooked the whole computer. The computer cost $237,500 dollars and must be replaced. The company's quarterly earnings statement will be severely impacted by the fact that down time on this high-speed machine is $168,000 per day.

At 5:00 A.M., the plant superintendent made a call to Great American Industrial Computer Company in Commerce City, California where it was 11:00 A.M. the previous morning. The GAIC people said they could put together another computer out of stock parts in inventory. Another paper machine in Canada would have the delivery of its new computer delayed a few days as a result.

GAIC can get the replacement at Los Angeles International Airport by 5:00 P.M. Los Angeles time. But it would take two days to get it to Sydney by the fastest possible air express service. So GAIC called Global Express and was told, "Get it to the Ajo Spaceport in Arizona by six o'clock today and we'll put it on the eight o'clock flight."

The fifty-pound assembly was loaded aboard a chartered Learjet at Long Beach at 4:00 P.M., flown to the spaceport, and put aboard the 8:00 P.M. Global Express flight. It arrived at Great Australia Spaceport 35 minutes later, or 1:30 P.M. local time there. Hustled aboard a waiting GAF Nomad, it was flown to the paper mill where it arrived and was put on line by 3:30 P.M.

The paper machine was off-line for 12 hours. Not counting the price of the replacement computer but figuring in the Global Express charges of $8,124.00, the company lost only $84,000 instead of the $336,000 it might have kissed goodbye if only air express was available. The company saved nearly a quarter of a millon dollars.

This scenario not only illustrates how low-tech industries like papermaking could take advantage of such a high-tech service as Global Express, but it also indicates that such a service is initially going to be useful in industrialized areas where down-times and "just-in-time" inventories are important in the competitive world of business. In brief, the first market areas are going to be industrialized ones.

Some of them are now relatively isolated from the rest of the world. Global Express holds the promise of alleviating this isolation. In fact, transportation isolation may be a factor that is retarding the growth of these regions.

Five industrialized areas were selected for this example:

Los Angeles—the southwest American industrial region.

New York—the northeast American industrial region.

Berlin—the central European industrial region.

Moscow—the east European and west Asian industrial region.

Tokyo—the east Asian industrial region.

Two industrial regions isolated by distance in the southern hemisphere were selected as well:

Sydney—the Australian industrial region.

Buenos Aires—the South American industrial region.

Just because the hubs are tagged with city names doesn't mean that the spaceports will be located in those cities as are present-day airports. For the purposes of this example, they merely tag a region. Suitable spaceport sites can be found within 300 miles of these named cities.

These seven are perhaps the primary industrial regions of the world. India and southeast Asia could have been included. However, additional regions would have complicated the example and contributed little to an understanding of the basic Global Express concept. Other regions were not excluded for any other reason.

The Great Circle Network of Global Express ground tracks is shown in Figure 23-2.

Most people don't understand great circle tracks because they visualize the Earth as a flat map rather than a globe. Even people who fly the New York–Tokyo route regularly may not realize why the Boeing 747 flies north out of New York and south into Tokyo.

Note that most of the ground tracks are over ocean areas or continental regions with low population densities. However, some extremely important exceptions exist.

The early years of spaceship operations to space and in the Global Express mode will encounter *perceived* safety objections.

"What if these rockets fall on our heads?"

Much of this has its origin in the fact that the early rockets were long-range artillery shells. The British, French, and Belgians still remember the German V-2, while the Saudis and Israelis have had more recent experience with the V-2's descendant, the Russian Scud.

In addition, objections can be anticipated from environmentalists who will oppose these overflights as well as most of the spaceport locations.

Figure 23-2 shows the location of what may be the potential safety sensitive areas of the early years of orbital and Global Express operations.

The strongest objections may come from the overflights in and out of the New York regional hub. These ground tracks cross the continental United States in an east-west orientation and extend up the northeast coast. Overflight objections may delay the eastern US spaceport hub.

Another sensitive region is Europe. A brief study indicates that a spaceport on the Iberian peninsula (Spain/Portugal) may alleviate some of the overflight safety sensitivity.

A much smaller safety-sensitive region exists over the middle of the Japanese islands. However, the Japanese may be willing to dismiss such concerns in their drive for industrial and market dominance.

The South American and Australian regions may not exhibit a high level of safety sensitivity because Global Express may be welcome. The great circle Global Express network is going to *end the isolation of the southern hemisphere and bring it into the expanding world market economy.*

Figure 23-3 is a table of great circle distances between the Global Express hubs of the previous two maps.

Potential great circle routes of 3,500 statute miles or less were not considered. At distances up to 3,500 miles, the air transit time is seven hours or less. These markets are already well served by overnight air cargo. A new business doesn't start successfully by going head-to-head with the giant in the marketplace. It chooses a market niche that isn't being served very well or one that the industry giants don't think is worth it.

Providing daily service between each of these seven hubs will require a large fleet of SSTOs. Assuming that one flight per day is made from each of the seven hubs to the other six hubs, this is 49 SSTO flights per day. Assuming a three-day turnaround time per vehicle, this will require 147 SSTOs. At a price of $500 million each, the capital investment in spaceships alone, leaving aside the spaceports and facilities, is $73.5 billion. If the all-important turnaround time can be cut to two days, this reduces the fleet to 98 ships at a capital cost of $49 billion.

Therefore, the service will start between two hubs at first—probably Los Angeles and Tokyo. This may build within a year or so

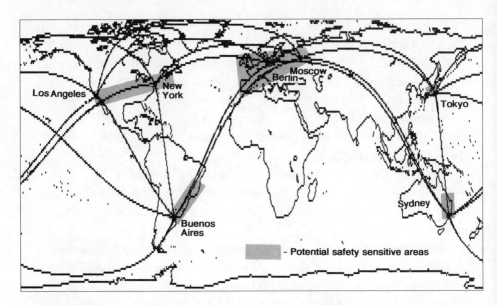

FIGURE 23-2: *Shaded areas of the great circle network of Global Express indicate potential safety-sensitive areas during the early years of the service. (Drawing by G. Harry Stine.)*

GREAT CIRCLE DISTANCES

(Applicable distances more than 3,500 miles.)

	Berlin	Buenos Aires	Los Angeles	Moscow	New York	Sydney	Tokyo
Berlin	-----	7,403	5,783	N/A	3,962	9,991	5,538
Buenos Aires	7,403	-----	6,113	8,378	5,299	7,321	11,397
Los Angeles	5,783	6,113	-----	6,071	N/A	7,497	5,474
Moscow	N/A	8,378	6,071	-----	4,664	8,997	4,554
New York	3,962	5,299	N/A	4,664	-----	9,931	6,734
Sydney	9,991	7,321	7,497	8,997	9,931	-----	4,858
Tokyo	5,538	11,397	5,474	4,544	6,734	4,858	-----

Copyright © 1994 by G. Harry Stine

FIGURE 23-3: *Great circle network distances between the seven industrial areas used as examples in the Global Express study. (Table by G. Harry Stine.)*

to four hubs, adding Sydney and Europe (Berlin). A four-hub Global Express system would require 48 SSTOs with three-day turn-arounds at a capital cost of $24 billion; cut the turnaround time to two days, and it requires 32 ships and $16 billion. At a per-flight cost of $1.5 million, the service would generate revenues of $8.76 billion per year.

Is the market there? The overnight air express companies think so right now.

It would be conceivable to start Global Express between two hubs with six SSTOs. In fact, this is the hypothetical company Paul Hans and I created for our economic model discussed in the previous chapter.

However, it may be too small. Federal Express started operations with a small fleet of French Dassault Falcon 20 bizjets pressed into cargo operations. A limited route structure was flown, eventually with up to 33 Falcons. Federal Express lost money. But Fred Smith proved his point, was able to acquire additional capital, and then greatly expanded the payload capability of his fleet aircraft and established a nationwide network.

Global Express can be a viable near-term factor in an SSTO spaceship operation that can help pay the bills while the Earth-orbit market is developed. It fits into the existing transportation systems and can begin operating as an adjunct to them, providing a niche service. It can provide operational experience and result in spaceports being built in several parts of the industrialized world. The existence of spaceports means a market for SSTOs.

Although the United States is in a favorable geographical position in a potential Global Express great circle network, it may turn out in the long run that Earth-to-Earth distances no longer mean anything but being first-to-market does. The quick interest of the Europeans and the Japanese in SSTO technology should now be a bit more clear.

To some people, this example may be superficial. I agree it's only a first approximation. Better data are needed.

Fast package delivery was part of the previously cited Commercial Space Transportation Study (CSTS) conducted by the Alliance of six major aerospace companies. I had no input to this study and never saw it until I received the final report in July 1994. I was encouraged that the study confirmed everything I've discussed here. What immediately drew my attention was the statement in CSTS regarding fast package delivery services:

> *Anecdotally and notably, interviewed personnel within the express package industry did not seem as skeptical or concerned as those in aerospace manufacturing with the makeup of the payload and the likelihood of a market materializing.*

The "makeup of the payload" refers to matters discussed in the previous chapter about the sorts of air cargo payloads that go into and out of the New York City airports.

The Global Express concept is a serious contender in the potential markets for SSTOs. It needs to be studied further while the SSTOs are being developed. Perhaps it can help provide some motivation to organize the new spacelines. However, a word of caution is advised here: It should not be studied to distraction while someone else is out there doing it. It looks like it's going to be a part of the growing global economy because a niche exists for the services it can provide. However, I hope the shipping invoices will be payable in dollars rather than yen or eurodollars. Studying the Global Express concept doesn't create those revenues. Only flying cargo and people from place to place will do that.

Americans have showed in this century that they're the world's best commercial airmen. It remains to be seen if they can seize such opportunities as this and become the world's best commercial spacemen as well.

TWENTY-FOUR

The Unbelievable Market

PEOPLE ARE LIVING AND working in space as you read this. Although it's true that since Skylab in 1973 they've gotten there at great expense riding atop ammunition, these pioneers proved it could be done. They were able to do useful work in space. And as Skylab, the Russian space station Mir, and space shuttle pictures show, they had fun, too.

However, when space tourism is mentioned, the first reaction from most people is mildly derisive laughter. Space tourism is an unbelievable market. However, disbelief doesn't mean that it's impossible or even unprofitable. With economical, reliable, regular space access, it becomes both possible and profitable.

I'm proud to be one of the first advocates of space tourism. Because of my 1975 book, *The Third Industrial Revolution*, I was one of the consultants on a 1976–1979 NASA study of space industrialization where my market studies showed that space tourism would become an economically viable market comparable in size to terrestrial ocean-going ship travel—i.e., 50,000 passengers or more per year—when ticket costs for a round-trip to orbit dropped below $50,000 per person.

I wasn't alone. The market potential for space tourism was independently evaluated by an organization no less prestigious than the Hudson Institute. Herman Kahn and William M. Brown authored a report entitled, "The Next 200 Years in Space," (HI-2352-RR, October 23, 1975, prepared for NASA Contract NASW-289).

The Hudson Institute forecast that space tourism would be one of the major economic areas of commercial space activities. Based on their estimates of costs, the estimate of the space tourism market was 10,000 people per year by 2005. The report concluded that space tourism would become "the single largest industry in space" in the next century.

The next stake in the ground was a paper by Society Expeditions presented at the Space Development Conference in Washington, D.C., on April 26, 1985. It stated that when seat costs dropped to $25,000 or less, the market would be 30,000 to 40,000 people per year.

This was being forecast when manned space flight involved shooting people into orbit on top of long-range artillery shells or variations thereof. Given the extreme cost of flying astronauts and mission specialists in the space shuttle, NASA required (they still do) people in outstanding physical condition who'd had at least six months of training. It was no wonder that the mere idea of space tourism was and continued to be met with disbelief if not derision because most people are aware only of the "NASA way" of putting people in space.

Space tourism cannot become reality as long as human travel in space is done the "NASA way." Little old ladies wanting to make ocean cruises do not book passage on torpedoes. People wishing to fly to Europe or Japan don't have to pass physical exams and undergo safety training in ground simulators. Even when it comes to understanding emergency procedures, no airline passenger has to be trained in an altitude chamber in case the airplane loses cabin pressure. Nor do they have to be trained to use the emergency exits or slides, life jackets, life rafts, and other safety equipment. In fact, most people don't even bother to read the safety brochure in the pocket of the seat ahead of them. The verbal briefing by the cabin crew is rarely listened to, being given only because FAA regulations require it. Yet history has shown that passengers use this equipment properly and even get out of the airplane safely in actual emergencies.

Much of the skepticism about space tourism stems from a confusion between "travel" and "tourism." Pulitzer Prize author Daniel

J. Boorstin pointed this out in his essay, "From Traveler To Tourist," published in his book, *Hidden History* (New York, Vintage Books, 1989, ISBN 0-679-72223-8). Before the 19th century, travel was a laborious, troublesome, and dangerous activity done only with the goal of high expectations in the form of profit. A *traveler* expected life-threatening danger, high adventure, and unanticipated experiences. Ocean-going sailing ship travel was originally the dangerous "travail" from which the word "travel" itself derives. With the introduction of the steamship and the passenger railway (a transportation system originally intended to haul coal), travel became a packaged commodity. This created the *tourist* who passively expected to encounter no danger, to live through the travel experience, to have things go as planned, and to have all the little details handled by the travel agent. A tourist goes "sight-seeing," a word coined in 1847.

Jules Verne's fictional Phileas Fogg went around the world in 80 days precisely at the time when travelers were becoming tourists. People were also growing richer because of the compound-interest principle of accumulation of capital and information. They could afford to take vacations by going somewhere. Verne's story was an excellent marketing tool for entrepreneurs who figured out how to make money providing services for the new tourists. Thomas Cook and American Express got their starts by packaging travel using the convenience and safety of the new railways and steamships, making reservations and issuing coupons for hotel stays, and selling traveler's checks that were convertible into the local currency everywhere.

After World War II when military operations showed how easy, cheap, and safe it was to fly across the oceans, steamship travel nearly sank into oblivion. The great ocean liners were scrapped or tied up to become theme restaurants. Then came a television series, "The Love Boat," that rejuvenated the ocean cruise by promising sex and adventure. It rescued the cruise ship and saved an industry. Some people made money and other people had fun.

Boat, rail, and even highway travel is used by a growing number of people from all walks of life for all sorts of business purposes, too. In 1830, anyone making a market analysis of how they thought people would use the new railways would have missed nearly all the reasons why people *did* ride the trains. Many aerospace engineers today appear to be missing the reasons why people want to fly into space.

Dr. Jerry Pournelle pointed out in his March 16, 1995, testimony before the Subcommittee on Space and Aeronautics of the House Committee on Science:

> Suppose in 1920 the Congress tried to form an intelligent estimate of the economic potential of airline travel; in particular the number of tickets that might be sold. One probable route would be New York City to Los Angeles, California. They might look at the number of people taking the trip by train. They'd then factor in the ease of travel by air as opposed to trains and try to guess at a number. If they felt bold, they might decide that as many as 500 a week would take the trip. Then they could be extravagant and multiply that by two to get 1,000 a week. They might even go mad and estimate 10,000 a week. . . . They'd never come close to the actual numbers on any reasonable or even sane set of assumptions, and even if they went mad and guessed the right numbers, no one would believe them, and they'd still not have a handle on the second order effects: the industries that are only made possible by rapid travel capabilities.

Tourism is a primary industry of a post-industrial culture. Kahn recognized this in his 1975 NASA study. The World Tourism and Travel Council in Brussels, Belgium, reported in 1995 that tourism accounts for 6% of the world's Gross Domestic Product.

Because of the ability to build reusable SSTO spaceships that operate like commercial aircraft, space tourism becomes possible if service is offered on short demand, costs and pricing structures allow a reasonable profit margin, and operations are conducted with the same level of reliability and safety as airlines.

However, is the market for space tourism truly unbelievable?

The best way to approach this is to look for an analogous market. In 1976, I chose ocean cruises although I could have gotten the same results by using the markets for long-range transoceanic first-class air travel.

A flight to orbit and back is roughly equivalent in terms of energy—but not time—required to make an around-the-world ocean cruise. How many people do this every year? In fact, how many people take ocean cruises that, in the aggregate, have similar total costs?

When I updated my 1977 space tourism article for the Summer 1990 issue of *The Journal of Practical Applications in Space*, I got then-current data from the Cruise Lines International Association of New York City. In 1989, 3,285,000 Americans took an ocean-going cruise, spending $4 billion for their ocean-going travel alone. This means that the average cruise cost $1,217.66 per person in 1989 dollars. With a multiplier factor of five brought in to determine total costs per person for the cruise—travel to and from the seaport, clothing, souvenirs, money spent ashore at ports of call, money spent aboard ship, etc., this brings the average cost per person for a modern ocean-going cruise to $7,305.96.

But the cost of going around the world on the Cunard liner H.M.S. *Queen Elizabeth II* runs from $15,000 to $100,000, depending upon desired accommodations. Several thousand people per year make equivalent trips on the *QE2* and other ships.

A round-trip coach airline ticket from Los Angeles to Sydney, Australia, costs about $3,600, according to the Official Airline Guide. First class is roughly double that or $7,200. At least three airlines fly the route with subsonic jet airliners on a daily basis—Qantas, Continental, and United. In 1986, they generated 8.3 billion seat miles at an average load factor of 73%. The distance between Los Angeles and Sydney is 7,480 statute miles. This means that 810,000 people flew between these points. Assuming that these were round trips, 405,000 people were carried in each direction. This route generates more than $2.9 billion in annual revenues.

True "adventure travel" is yet another analogous market. If you want to climb Mount Everest—first conquered in 1953 at great expense and effort—it will cost $50,000 and there's a waiting list.

Thus analogous markets for space tourism exist. Two of them are large. The largest in terms of both dollars and number of people is the ocean cruise market. Both indicate that a space tourism market can be roughly in the same ballpark, and this ballpark is large.

This was well understood by Society Expeditions of Seattle, Washington, as discussed earlier. To recap, in the early 1980s Society Expeditions showed that a market for space tourism existed. They offered a one-day trip to orbit for a ticket price of $50,000. But they couldn't buy or lease a space shuttle from NASA, which turned out to be a fortunate bit of bad luck for them. Society Expeditions was in the process of raising the capital necessary to purchase their own Phoenix SSTOs from Pacific American Launch Systems when the space shuttle Challenger blew up in 1986.

Society Expeditions returned more than 500 cash deposits of $5000 they'd been holding in escrow.

Thus, a viable space tourism market exists, given suitable low-cost, reliable, safe, scheduled space transportation.

But what would a ticket to orbit cost?

Referring to the economic space transportation computer model discussed previously: At a per-flight cost of $1.6 million for an SSTO configured to carry 110 passengers into orbit and back, the ticket price would be:

$$\$14,600.00$$

This is certainly within the market price range of the existing tourism analogs and less than one-third the $50,000 ticket price envisioned by Society Expeditions, Inc.

Would people be willing to buy tickets to orbit at $14,600 each? Would you?

This speculation was further confirmed by the previously cited 1994 Commercial Space Transportation Study. It's highly conservative in most respects. So if you read the CSTS report, please keep this in mind. However, it added a data point: William Buckley organizes an annual around-the-world trip by air with various stopovers at a cost of between $60,000 and $80,000 per person. The trip is always booked solid well in advance.

What sort of orbital trips can we expect to see in the near future?

One of these is the "joy ride," a suborbital ride in a Global Express flight or for a duration of a few orbits.

The next is the orbital visit to a simple orbital facility for durations of three to ten days. Passengers would debark the SSTO at the station and the spaceship would be sent back to Earth, probably with another load of tourists whose time in orbit was up. It doesn't pay to put people in orbit and keep them in the spaceship for three days. In the first place, the spaceship would be tied up on that operation for that time period, and it wouldn't be flying with other revenue payloads. Hence the ticket price would have to be increased.

Besides, who flies to Hawaii, stays in the 747 for three days, then flies home again? Living in a spaceship will be no fun after the first few hours. It will be like living inside a travel trailer for an entire vacation and not being able to open the door.

Obviously, someone is going to build a space hotel for the space tourists. Hilton? Marriott? Hyatt? The first one will be busy! And it may well become the first real space station in the process because economics will drive its establishment.

What are tourists going to do up there?

Some will go just to get into space.

Some will go to watch and wonder at the view of the Earth unfolding through the porthole in great and glorious detail. A television anchorman once asked me, "Space tourism? Who wants to see the Grand Canyon from a hundred miles away?" I told him I did because I'd never seen it from a hundred miles in space; anyone can see it from the ground.

Other people will experiment with what has been called "humanity's favorite recreation."

Astronauts are always asked two questions. One is, "How do you go to the bathroom in space?" The late Ron Evans used to give a humorous lecture about this entitled, "There Ain't No Graceful Way." The other question is asked in private: "How about sex in space?"

This appears to be a very strong motivation behind the desire of many people to go into space.

It's a good thing, too. People will have to mate in space if they intend to colonize the solar system. If they don't, we'll never be a spacefaring people. So in the minds of the prudes, we're damned if we do; in reality, we're doomed if we don't.

Several years ago, some clandestine experiments were conducted late at night in the neutral buoyancy weightless simulation tank at NASA's Marshall Space Flight Center in Huntsville, Alabama. This huge swimming pool allows astronauts to simulate zero-gravity by working under water. It's amazingly realistic. (Try changing the light bulb in a swimming pool while staying underwater with a snorkel or scuba gear.) Although the scuba gear tended to get in the way at Huntsville, the report came to me in a plain brown telephone call. I was told that, yes, it is indeed possible for humans to copulate in weightlessness. However, they have trouble staying together. The covert researchers discovered that it helps to have a third person to push at the right time in the right place. The researchers—who must remain anonymous lest they get into trouble—discovered that this is the way dolphins do it. A third dolphin is always present during the mating process. Therefore, the space counterpart of the infamous aviation Mile High Club is the Three Dolphin Club.

I wrote a science fact article about this and got a lot of mail, including a report from people who worked for Lockheed at the NASA Kennedy Space Center in Florida. They included a drawing of the membership pin of the Three Dolphin Club, shown here as Figure 24-2. As of 1990, several space shuttle astronauts were qualified to wear this pin because of nonscheduled personal activities aboard the space shuttle on *seven flights*. I shall protect the names of these pioneers of human expansion into space because I don't want them to lose their astronaut status.

The Japanese Rocket Society devoted the entire Spring 1993 issue of their *Journal of Space Technology and Science* to space tourism. It included an article by Toyohiro Akiyama entitled, "The Pleasure of Spaceflight." Another Japanese journal, *Space Engineering/Construction/Operations*, recently carried two space tourism articles covering the potential demand for passenger travel to orbit and the design and construction of zero-gravity sports centers.

The Japanese have caught on. Being of a different culture that's somewhat less puritanical but in a different way than that in Amer-

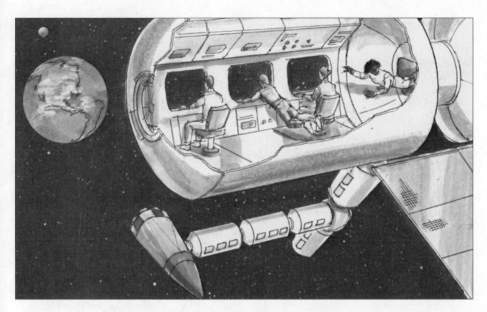

FIGURE 24-1: *Space tourism also means space hotels. Here a group of astro-tourists wait while their SSTO approaches and docks to the space hotel. (Courtesy Dr. William A. Gaubatz, McDonnell Douglas Corporation.)*

FIGURE 24-2: *The "official" insignia of the Three Dolphin Club sent to the author from nameless contractor employees at NASA Kennedy Space Center. (Courtesy of protected sources at Cape Canaveral.)*

ica, representatives from Shimizu Corporation and others frankly admit that they are quite serious about starting space tourism at the earliest possible opportunity.

They've already identified their primary target market: honeymoon couples.

In 1995, researchers from Tokyo University, the Japanese National Aerospace Laboratory, and the York Business School in Toronto, Canada, published the results of an in-depth space tourism market study conducted with grants from such Japanese firms as Shimizu and Mitsubishi Heavy Industries. They surveyed 3,030 people in Japan in 1993 and 1,020 people in the United States and Canada in September 1995. 70% of the Japanese respondents and 60% of the American/Canadian participants said they were interested in traveling in space for a vacation. The study revealed that if the ticket price was between $15,000 and $50,000, the estimated space tourism market within ten years of initiation would be one million passengers per year, generating revenues of $10 billion or more.

This study revealed the design of a Japanese reusable SSTO, the "Kankoh-maru."

In late 1995, NASA and the Space Transportation Association initiated a space tourism study to further expand those of the

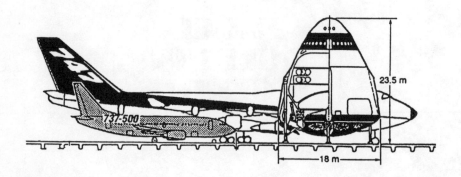

FIGURE 24-3: *The proposed Japanese "Kankoh-maru" passenger-carrying SSTO compared in size to a Boeing 747-400 and a Boeing 737-500. (From Nagatomo et al, paper before 6th International Space Conference on Pacific-Basin Societies, December 1995.)*

Japanese and the 1994 Commercial Space Transportation Study. I was interviewed at length by Dr. Barbara Stone of NASA's Office of Space Access and Technology and also serve on the study's steering group.

Space tourism won't continue to be an unbelievable market when it dawns on people that they don't have to be an astronaut to travel in space and even more so when they realize they'll be able to afford it. At less than $15,000 per ticket, probably with low-rate APR financing available, they won't buy a new car this year but a ticket to go into space instead. There they'll find exactly what tourists have always wanted: adventure, seeing and doing something new, and coming home to talk about it.

TWENTY–FIVE

Spaceports

SSTO SPACESHIPS WILL REQUIRE spaceports just as airliners require airports. A spaceport will be a place where spaceships take off to fly into space and land after flying back from space.

A spaceport also will be a place where spaceships are manufactured, tested, maintained, modified, and repaired. Although airplanes are flown from the factories where they're built, we certainly can't expect McDonnell Douglas, for example, to get the FAA and the Los Angeles area authorities to let Delta Clippers be flown to their new owners from the Huntington Beach plant where MDC expendable space vehicles are built and shipped elsewhere by truck. A 50-ton SSTO 150 feet long isn't easily shipped by truck or rail. In fact, it isn't shipped any more than an ocean liner is transferred from nearby Long Beach harbor overland to a buyer in Houston. The old Apollo Saturn S-IVB stages, the *smallest* of the moon rocket stages, had to be shipped by barge from nearby Seal Beach through the Panama Canal to Cape Canaveral.

Cargos will be received at the spaceport from other transportation modes—rail, highway, and air—where they'll be processed and loaded aboard spaceships for delivery elsewhere in the world by Global Express or to orbit. Some cargos will be returned from orbit by spaceships and then transshipped to their ultimate ground destinations by rail, road, or air.

As space tourism grows, a spaceport must be capable of handling people, catering to their needs and ensuring that they can get to and from the spaceport.

A spaceport also means people to perform all the functions necessary to operate a major node in a transportation system. To get some idea of what this amounts to, spend some time at a major metropolitan airport. Some people refuel aircraft. Others are cargo handlers. Some are specialists in warehousing. Yet others provide security, collect parking lot fees, provide food for other people who work there and for passengers, take care of utilities and sanitation, and—in short—carry out all the functions necessary to run a small city. Some airports like Los Angeles International, Chicago O'Hare International, and John F. Kennedy International *are* cities all to themselves.

Back in the days when the railroads were the primary haulers of people and cargo, it was said that you could spend your entire life inside New York's Grand Central Terminal and never lack for any amenities of life. You could literally live inside the terminal complex. The same holds true today with major airports. Spaceports will be no different.

Initially, a spaceport may need operating facilities that will handle spaceships that operate in any of three modes: VTOVL (Vertical TakeOff Vertical Landing), VTOHL (Vertical TakeOff Horizontal Landing), and HTOHL (Horizontal TakeOff Horizontal Landing). This means both concrete pads and runways.

One operating mode will turn out to be the most economical, safe, reliable, and profitable *at the time.* This can and probably will change as technology progresses. At this early point in the commercial space industry, don't make the mistake of betting too heavily on existing technology. It may get us there and get us back, but new technology will come along. The zeppelin companies and the flying boat builders learned this the hard way. One operating mode may indeed dominate the scene at the start because it's simpler. Given the proven ability of vertical takeoff and landing spaceship to recover from an emergency during launch, this mode may continue to lead.

The spaceship operators will stay away from the existing national launch ranges—Cape Canaveral, Vandenberg Air Force Base, White Sands, and Wallops Island. These places have operational rules steeped in the artillery shell activities of the past. SSTOs need none of these.

Because SSTO spaceships will be designed for total reuse (no dropping of parts), engine-out capability (ability to recover from a partial propulsion failure), and flight safety (redundant or back-up systems such as are already used in all airplanes, including airliners), they no longer have to operate from artillery ranges like expendable launch vehicles.

The true spaceports cannot afford the delays and other operational procedures of these super artillery ranges. Given the aircraft-like operational characteristics of reusable SSTO spaceships, a spaceport can therefore be located nearly anywhere. However, spaceport sites will have to meet specific physical and economical criteria.

A spaceport must have the following:

1. Transportation access by rail, highway, and air.

A spaceport sited where no one or nothing can get to it easily is useless. There's only one known bridge that wasn't built to handle traffic, much less handle it better: the tourist attraction across the Royal Gorge in Colorado. Spaceports must be integrated into the existing transportation networks so they can take maximum advantage of the flexibility this creates. Isolated transportation hubs have never succeeded. Transportation hubs, including spaceports, must be accessible. Rail access is mandatory because most heavy, bulky loads are transported by rail. Rail access will reduce the cost of building and modifying large spaceport structures, for example. Highway access is also imperative because 18-wheeler semitrailer trucks can take smaller loads nearly anywhere that roads exist. Air access is also required for speedy delivery and pickup of time-critical cargos and people.

2. Clear airspace corridors for departures and arrivals of all types of spaceships.

Present-day space launch sites patterned after artillery ranges require buffered Restricted Areas to reduce the possibility of interference between space launches and commercial air traffic. Although the chance of a midair collision between a spaceship and an airliner will be infinitismally small, the *perception* of hazard isn't. In the early years, airspace use conflicts around spaceports may pose real problems for commercial SSTO operations. It's expensive for an SSTO to sit on the ground boiling off its cryogenic propellants while it waits for a Boeing 777 to pass through its departure airspace. And it's extremely difficult to put a returning spaceship into a holding pattern while an airliner clears the landing corridor's airspace.

So spaceports must have the commercial airspace around them designed to provide clear approach and departure corridors. Or they will be sited where the existing airspace can accommodate such a requirement. Such latter sites are hard to find in the industrialized parts of the world where the majority of Global Express and space traffic will exist. As time goes by and more experience is gained in the mix of air traffic and space traffic, the airspace requirements will be relaxed. But not at first. The author has had extensive experience dealing with the FAA on airspace issues, and caution is the order of the day. It's *not* a trivial problem. But solutions have been developed, even if they have no official recognition yet.

3. Proximity to existing natural gas pipelines or coal reserves.

Natural gas or coal are excellent sources for hydrogen which, in its liquified form, is likely to be the most common SSTO rocket fuel in combination with liquid oxygen as the oxidizer. Even air-breathing spaceplanes, if they become feasible, will probably require liquid hydrogen because it's one of the most energetic and environmentally benign of all fuels.

4. Proximity to a high-capacity electric power grid.

Electricity is our most ubiquitous and easily handled energy source. It can be "piped" around in wires. It isn't even created until it's demanded. A spaceport will need electrical power, and the closer it is to existing sources and power lines, the less expensive the construction and operation of the spaceport.

Liquid hydrogen and liquid oxygen will have to be produced at the spaceport because of the quantities and the high daily flow rates that will be needed. Except for the very early years when use rates are low and the propellants can be brought in by truck or rail, it will be more economical to produce propellants on the spaceport.

Liquid oxygen can be obtained by liquefying and fractionally distilling ordinary air. It can also be obtained by the electrolytic decomposition of water—fresh, hard, or seawater can be used although the design of the electrolysis plant is different for different water types.

Liquid hydrogen can be obtained from fossil fuels such as coal, crude oil, or natural gas. It can also be obtained from the decomposition of water.

Obtaining cryogenics from any source requires electric energy.

Keeping cryogenics cool requires energy.

And the spaceport itself will need electricity. New York Kennedy, Chicago O'Hare, Atlanta Hartsfield, or Los Angeles International consume megawatts on a daily, round-the-clock basis. Megawatt coal, natural gas, or nuclear electric power plants are not cheap, nor are they built overnight. A spaceport will need to tap the existing regional electric power grid. The closer it is to a major electrical transmission grid, the cheaper it will be to build and operate it. New electrical generating systems such as fuel cells powered by natural gas show promise of eliminating the need to be near the electric power grid.

5. A "noise clear" zone extending from five to fifteen miles around the spaceport.

Rockets, injected ramjets, and other jet propulsion engines do not operate quietly. Many years will pass before they are quieted because reducing the sound levels usually means reducing the propulsive efficiency of the engines. In time, the noise will be ameliorated as technology advances. But spaceports at first will be noisy places.

Vertical takeoff spaceships have a circular noise footprint and may require a small takeoff noise buffer zone.

Horizontal takeoff spaceships have a noise footprint similar to that of an F-15 or F-22 taking off with afterburners operating.

A vertically-landing spaceship can use the same sort of cylindrical airspace corridor it requires during its vertical departure. This will probably be five miles in diameter extending from the surface to 60,000 feet, the highest altitude over which the FAA has jurisdiction.

A landing noise corridor will be required because of the sonic boom associated with horizontal landing. The NASA space shuttle

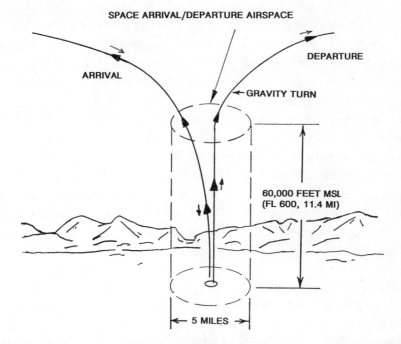

SPACE ARRIVAL/DEPARTURE AIRSPACE

DEPARTURE

ARRIVAL

← GRAVITY TURN

60,000 FEET MSL
(FL 600, 11.4 MI)

|← 5 MILES →|

FIGURE 25-1: *Drawing depicting the airspace "bucket" five miles in diameter and 60,000 feet high for VTOVL spaceships flying in FAA-controlled airspace. (Drawing by G. Harry Stine.)*

Orbiter produces a sharp double bang during entry. Sonic booms don't have to be harmful. And they don't have to be startling if the 10:15 arrival is expected every day.

This "noise clear" requirement means that the early spaceports can be close to but not in major urban areas.

However, *remote* locations for spaceports are no longer necessary for artillery range safety purposes, which is why White Sands and later the Cape were sited as they were.

Most places in the world will meet all but perhaps one of those five spaceport criteria. So will spaceports be built there? Not necessarily. The early ones will be built where the customers for Global Express and space transportation exist.

Therefore, the new SSTO spaceships will no longer be chained to existing expendable rocket launch sites. A brief look at the historical background will show why.

The original requirements for a site to launch large rockets were stated by Colonel Walter R. Dornberger in Germany in 1935. He was an artillery officer who had been put in charge of the secret German project to develop a long-range artillery rocket with twice the range of the legendary Paris Gun. This rocket became the German V-2, the ancestor of all present-day expendable space launch vehicles.

Dornberger wrote that a rocket launching center should be located on a seacoast because Germany didn't have open land on which these artillery rockets could crash. He recommended a sea-coast site so the line of flight could be parallel to a long stretch of coastline. Since the German rocket program was secret and a lot of rockets blew up in those days, he wanted the facility to be on flat land in a remote location for security and safety.

Dornberger described the ultimate firing range for long-range artillery.

We're still wedded to Dornberger's requirements because present expendable space launch vehicles were developed from long-range ballistic missiles (ICBMs), the descendants of the V-2.

Every space launch center in the world is nothing more than a super artillery range. The differences between space launch centers and airports—Cape Canaveral versus Orlando International Airport, for example—are striking.

But some misperceptions will have to be surmounted first:

"What if these spaceships fall on our heads?"

People asked the same question about airplanes in the 1920s.

The safety concern of flying spaceships over populated areas disappears with the advent of reusable SSTO spaceships that will not drop boosters or lower stages in flight. They will have engine-out capabilities and safe abort modes similar to airplanes. How often does a jet airliner lose all its engines leaving San Francisco International?

The prohibition against flying rockets over populated areas arose on May 29, 1947 when a German V-2 flew off the White Sands range in New Mexico and landed in Ciudad Juarez, Mexico. The present day flight safety system for ballistic missile and space vehicle launch sites evolved from the procedures put in place immediately thereafter by an artillery test officer in the U.S. Army Ordnance Corps, Major Herbert L. Karsch.

But by 1956, Karsch and others at White Sands—including me—were working on a proposal to launch long-range ballistic missiles northwest from White Sands with impact areas in Utah, the northwestern United States, and Alaska. This was never done because Cape Canaveral was a better artillery range for ICBMs. But missiles were later launched from Green River, Utah, to impact on the White Sands range. Rocket flight beyond range boundaries and over populated areas became accepted in specific instances.

With SSTO spaceships, such flight restrictions are no longer necessary. The new spaceports will be located where they're needed, just like airports.

What will a true spaceport look like?

It won't resemble today's space launch sites. There will be no massive mounds of solid concrete with their canyonlike flame trenches into which millions of gallons of water are released just before launch. Gone will be the towering steel structures of mobile skyscrapers needed to service expendable launch vehicles. There's no need for massive concrete and steel blockhouses for hundreds of people on the launch crew. SSTOs won't need the huge hangars where today thousands of people assemble an expendable launch vehicle and its cargo for months before flight takes place.

A commercial spaceport will resemble an airport. In fact, it will also operate as an airport capable of handling space passengers and cargo arriving and departing by air.

A spaceport will consist of concrete pads from which the SSTO spaceships depart and concrete pads where they land. If the spaceships are horizontal landers, they'll touch down on the same runway used by the airliners and cargo planes that fly in and out of the spaceport. The pads will be connected by ordinary roadways along which the SSTO spaceships can be towed on their landing gear to move them from the landing pad to the flight pad or, if they need maintenance and repair, to the hangars that are, at most, 15 to 20 stories high. Propellant manufacturing and storage facilities will be located reasonably near the launch pads. The other buildings on the spaceport will look like industrial buildings everywhere. Because noise level is perhaps the biggest problem, they may be several thousand feet away from the launch and

FIGURE 25-2: *Spaceports for SSTOs will resemble airports and will consist of launch and landing pads connected by roads to hangars. (Courtesy Dr. William A. Gaubatz, McDonnell Douglas Corporation.)*

landing pads, and their windows may be smaller with thicker panes of polycarbonate material that is less brittle and stronger than ordinary glass.

"But won't the distances have to be greater? Think of all that explosive rocket fuel! What happens if a spaceship blows up?"

This is "dinosaur thinking" left over from ballistic missile days. The chances of an SSTO spaceship blowing up are about the same as a 747 blowing up on takeoff. Remember: These spaceships will be designed and operated like airplanes. Rocket propellants such as liquid oxygen and liquid hydrogen have been handled safely for more than a third of a century. Both of these are also used in million-gallon quantities by industry every day. Handling them is no more dangerous than handling gasoline. The National Fire Protection Association (NFPA), a nationwide voluntary standards organization, has standards and codes already written and in use for storing and handling these cryogenic liquids.

A spaceport will even be operated like an airport when it comes to spaceships taking off and landing.

After all, we're entering a new era of commercial space that has shucked off the needless trappings of the missile age.

As of this writing, several communities are planning spaceports.

The New Mexico state government has hired people to plan the Southwest Regional Spaceport west of White Sands Missile Range.

Florida has established a commercial spaceport site on its Atlantic coast. Florida may run into problems with weather such as hurricanes. And the humid climate has never been one to ameliorate problems of corrosion.

Space enthusiasts in Colorado have proposed both military and commercial spaceports on the plains east of Denver. But these plans may run into airspace problems created by the new Denver International Airport. Weather, mostly snow, is also a problem.

Alaska has opened a commercial spaceport intended for launches into polar orbit. But it may suffer from being too far from the industrial portions of the world that can use its services, and it's too far away from the established air, rail, and highway networks that must feed it.

Other states are entering the lists even as this is being written. An Aerospace States Association (ASA) is already in existence, its membership made up of those states that have space commissions or governmental agencies involved with promoting future economic growth.

I'm a member of the thirteen-person Arizona Space Commission. We're looking at *several* potential spaceport sites in Arizona, especially those in close proximity to Mexico because, with the North American Free Trade Agreement (NAFTA), Mexico can become a major player in the commercial space age, too. In the meantime, Arizona is a member of the ASA and cooperating with New Mexico in the Southwest Regional Spaceport.

However, to fixate on the idea that there will be one and only one commercial spaceport is as shortsighted as believing that there will be one and only one national airport. With low-cost, economical, reliable, responsive, on-demand space transportation available within ten years using reusable SSTO spaceships, there will be *many* spaceports if initially only for Global Express operations. Count on Japan having its own spaceport because in mid-1994 the Japanese announced that their future space program would involve reusable SSTO vehicles in spite of the fact that they'd just flown their first expendable H-2 rocket.

If the whole commercial space affair was nothing more than a bunch of space advocates talking about SSTO and NASA doing SSTO studies and maybe flying a few X-vehicles, the new commercial space age might be dismissed as another techie dream. However, when investment firms, state governments, and even other countries start *doing things* instead of talking about them, something real is happening.

With spaceports and spaceships, we'll have space transportation for everyone, right?

Wrong!

Don't forget the politicians, bureaucrats, and lawyers.

If you neglect to consider them, we'll continue to face the problem that we can't get there from here.

TWENTY–SIX

"Superata Tellus Sidera Donat"

"Overcome the earth and the stars shall be yours," said the Roman philosopher Anicius Manlius Severinus Boethius (480?–524?) in his "Consolatio Philosophiae." His words have special meaning today when applied to commercial space transportation.

In the next ten years, we can be halfway to anywhere in the solar system. No scientific or technical breakthroughs are needed to build spaceships that operate like airplanes. Applying this technology is a matter of grungy, grinding, grutty engineering—building and busting X-vehicles, dealing with the little failures of design that always accompany any project, and kicking technology in the teeth hard enough and often enough to get it to do what is wanted.

The economics support it. Spaceships built and operated like airplanes can be at least two orders of magnitude less costly to build and operate than the expendable disintegrating totem poles we've used since 1957. It's possible to make money with these reusable spaceships.

Some of the markets for reusable SSTOs exist while others can be forecast. Yet greatly reduced space transportation costs will cause such a nonlinear revolutionary change in the existing markets that it's difficult to nail them down with the sort of precision and thoroughness demanded by today's financiers with their computer spreadsheets. Space transportation always will be "expensive" in the minds of some people, but the word is a relative term. It's also perceived as "risky," but those who lead in business and

industry are the ones who discover risk is manageable before anyone else comes to that conclusion.

However, even if we have the technology and the money to put
it to use in developing commercial space transportation, we may
not be permitted to do so. The way to space may be blocked by
lawyers, politicians, generals, and bureaucrats.

*Regulatory and liability issues may freeze commercial space
activities in their tracks.*

In the classic 1950 motion picture, "Destination Moon," the
county sheriff shows up at the last moment with a court injunc-
tion to halt the launch of the single-stage spaceship *Luna*. It's not
out of the question even today (provided that the situation could
even get as far as an impending launch before the activity is
brought to a halt).

So take these nontechnical and non-economic issues to heart.
Do not automatically assume that someone else is going to take care
of them. They may not. And these issues can stop a space
entrepreneur as surely as the Federal Aviation Administration can
keep people from building and flying their own airplanes without
approval. People who want to become space entrepreneurs must get
busy, solve the nontechnical problems, correct the situations, or
change the treaties, laws, statutes, rules, and regulations *now*
because it will take five to ten years to get results.

The tort liability issue is one matter that needs fixing.

At this time, if a Northwest or United airliner, for example,
leaves Tokyo's Narita International Airport and crashes in the Ginza
for whatever reason, the airline itself is responsible under the terms
of the *1952 Rome Convention on Damage Caused by Foreign Aircraft to Third Parties on the Surface*. The airplane may be registered
in the United States, but the airline will be held responsible and the
airline's insurance carrier will have to handle the damages and lawsuits.

However, if any sort of space vehicle has a similar accident in
which property is damaged or lives lost, the incident is presently
covered by the *1972 Convention on International Liability for Damages Caused by Space Objects* (24 UST 2389, TIAS 7762). This
treaty is a special agreement going into greater detail than the *1967
United Nations Treaty on Principles Governing the Activities of*

States in the Exploration and Use of Outer Space, Including the Moon and Other Celestial Bodies. The United States is a party to these as well as two other conventions that arose from the more general 1967 Treaty of Principles. But it is the 1972 International Liability Convention that's more than bothersome.

Under the provisions of this early-day space treaty, if a civil spaceline operator's spaceship has an accident, the company isn't liable. *The government of the country where the spaceship is registered is individually and severally liable under the terms of that treaty.*

The 1967 Treaty of Principles and the 1972 Convention on Liability are dinosaurs from the days when only expendable launch vehicles existed and only governments were involved in space activities. The basic Treaty of Principles also is a vestige of the Cold War because it was originated by the Soviet Union and is strongly anti-free-enterprise. It requires that "all activities in space be conducted exclusively by states" and requires authorization and continuing supervision by the appropriate state party to the treaty of any nongovernmental entity operating in outer space (Article 6). This treaty is without parallel in international trade and commerce.

Thus the federal government had to establish the Office of Commercial Space Transportation (OCST) within the Department of Transportation because of this treaty.

The OCST in turn established a body of regulations. These are in the Code of Federal Regulations available at local libraries. The portions covering commercial space activities are set forth in 14 CFR Chapter III Parts 400 to 415 inclusive.

The regulations cover safety inspections, licenses, clearances, permits, and approvals for each individual space launch operation, regardless of the type of launch vehicle used.

OCST representatives have open authority to enter and inspect any and all facilities of a launch vehicle or payload builder or operator.

OCST can suspend any license they grant if they believe that the "national security or foreign policy interest of the United States" may require it [14 CFR Ch. III, Part 405.3(a)].

These regulations will be unworkable and unenforceable with frequent commercial SSTO spaceship operations because they were developed in response to the Commercial Space Act of 1984, passed by Congress as a result of the morass of laws, rules, and regulations that had to be followed by Space Services, Inc., in launching their privately-funded Conestoga rocket in September 1982.

These U.N. treaties are an academic exercise in international law made far in advance of the reality they purport to regulate. They fail to address the legitimate needs of private corporations and individuals who own space resources and exploit them for profit. The treaties are, in reality, political statements by the former Soviet Union and Third World nations more than they are a workable set of legal rules governing the actions of people developing and using space resources.

Therefore, the United States needs to investigate getting these treaties amended to reflect the reality of commercial SSTO spaceship operations. If this cannot be done, then the United States' ratification of these treaties should be revoked.

The people who wrote the United States Constitution were leery of international entanglements. So they adopted specific procedures relating to the adoption of treaties by the United States. Ratifying a treaty and thus becoming a party thereto requires the advice and consent of the United States Senate.

But they didn't include a constitutional procedure to withdraw from a ratified treaty that has become useless, obsolete, or harmful. This was worked out only in the 20th century. Abrogation or revocation of participation in a treaty by the United States requires only an Executive Order of the President. This happened when President Jimmy Carter revoked the Panama Canal Treaty by means of such an Executive Order. Former Senator Barry M. Goldwater (R, AZ) didn't think that was constitutional, so he sued the Carter administration in federal court. Senator Goldwater maintained that only the Senate can properly, under the Constitution, revoke a treaty because that is the body that advises and consents to the ratification of the treaty to begin with. Senator Goldwater lost. Therefore, this established the precedent that the United States can back out of any treaty by a presidential order.

Modifying or withdrawing from these treaties must be done to ensure the future of commercial space operations, now that we are on the verge of having the sort of space access that will make such operations possible and profitable. As Arthur M. Dula pointed out in a 1981 memorandum to Space Services, Inc., it's a matter of international law. He went on to say that the treaties didn't mean much to a majority of nations because the expense involved in space activities precluded private initiatives as a matter of economics, not international law. With the expense of space transportation being reduced by two orders of magnitude, it indeed becomes a matter of international law based on "customary law," international agreement, or derivation from general principles of law common to the major legal systems of the world.

However, in late 1995, the Office of Commercial Space Transportation was merged into the Federal Aviation Administration, also part of the Department of Transportation. The regulations of the two agencies, now merged, are in conflict.

Another problem arises when it comes to the matter of certification of "spaceworthiness" for spaceships under international law.

At this time, every airplane that crosses an international border must carry a certificate of airworthiness issued by the country of aircraft registry. It must also carry a certificate of manufacture and a cargo manifest.

However, this issue was considerably muddied in the case of the NASA space shuttle. Originally, Orbiters didn't carry these documents. This oversight was brought to the attention of NASA not by a memorandum but by the publication of a contemporary novel in which this was a major factor preventing the recovery of a space shuttle Orbiter that had made an emergency landing on foreign soil. Now each Orbiter carries all the necessary documentation except a certificate of airworthiness because the Federal Aviation Regulations have no provisions for certificating a space vehicle. In fact, the FAA legal counsel dodged the issue back in 1977 by ruling that the Orbiter wasn't an airplane even though it had wings and was a glider during its landing phase.

The legal problem of certification remains unsolved because no one could see the need to consider it. But this will have to change once privately-owned commercial SSTO spaceships begin flying to orbit and back. Therefore, effort needs to be given to drafting proposed regulations in this regard.

On the basis of precedent—something that attorneys rely upon heavily—no one knows if reusable, piloted, commercial SSTO spaceships capable of carrying people must be certificated under the provisions of the Federal Aviation Regulations or not. Certainly, OCST wasn't given this authority or responsibility under the provisions of the Commercial Space Launch Act of 1984 (Public Law 98-575, 49 U.S.C. App. 2601).

It's prudent to assume that such spaceship certification regulations and procedures will be written and implemented because a 1957 United Nations General Assembly Resolution states that air law applies to spacecraft operating in airspace. Therefore, any private reusable launch vehicle company should anticipate having to operate spaceships with airworthiness certification, especially if passengers are carried. There are several places to look for guidance concerning what these regulations and procedures may turn out to be.

In concert with the similar aviation agencies in other countries, FAA presently certificates *all* crewed aircraft under the provisions of Federal Aviation Regulations (FAR) Parts 23, 25, 27, 29, and 31. FAA accepts airworthiness certifications issued by other countries. The regulations governing airworthiness certification are comprehensive. They have created a level of aircraft safety unprecedented in the annals of transportation. But it costs money and takes one or more years to get FAA certification on a new airplane design.

It's easier to get an airworthiness certificate if the builder of the craft institutes a Reliability Centered Design and Maintenance program from the day the design work begins. The details of this procedure are spelled out in FAA Advisory Circular AC 120-17A. Basically, this involves keeping thorough and complete reliability records of every component, subsystem, and system that goes into the craft, even at the X-vehicle and prototype stages. Other data inputs are pilot reports, the results of functional checks, findings from sampling inspections, and service difficulty reports. Thus a track record of performance and reliability is established. Airplane manufacturers use this procedure because it shortens the certification period, and airlines use it to maintain reliable operation and reduce costs. For example, it was once mandated that certain jet engines had to be torn down and overhauled after a given number of operating hours. Using the reliability criteria, maintenance people track critical performance parameters instead so that when they see a component or subsystem begin to degrade, they know they'll have to perform repairs soon.

Perhaps FAR Part 30, Airworthiness Certification of Spaceships, will be developed out of the pertinent existing FARs.

Yet another legal and operational issue arises over the definition and establishment of boundaries for airspace and orbital space.

No accepted international definition of the boundary between airspace and orbital space exists. Thus, the question arises: Where does airspace end and where does orbital space begin? Or does it make any difference?

Indeed it does make a difference, especially from the standpoint of national security.

At the present time, the official upper limit of airspace over the United States is 60,000 feet above mean sea level (MSL). However, some military aircraft fly above that altitude. Under international law, the United States government has the right to permit or deny access into its national airspace from the earth's surface to 60,000 feet MSL.

Who has the authority over and therefore the responsibility for anything that flies above 60,000 feet MSL?

No one knows at this time.

Does outer space therefore begin at 60,001 feet MSL?

No.

The USAF awards astronaut wings to pilots who have flown above 50 miles (264,000 feet MSL) while the Fédération Aéronautique Internationale, the official international aeroclub with headquarters in Paris that homologates aviation and space records, says that space begins at 100 kilometers (62 miles). The Office of Commercial Space Transportation of the FAA hasn't officially addressed the problem.

Unofficially, it has, however. When the Notice of Proposed Rule-Making (NPRM) for the OCST regulations came out in April 1988, the agency proposed to require licenses, permits, inspections, and other regulatory controls for all rockets and satellites launched into space and into orbit as well as "all suborbital rockets." I made a telephone call to Washington and asked, "How do you intend to apply these regulations to the ten million suborbital model rockets launched by hobbyists and schools every year?" After a suitable period of consternation, they asked me to propose regulations that would exempt model and "amateur" rockets that didn't fly into space. But no one at OCST would agree on an altitude at which space begins. They didn't have that authority. So I proposed the 100-kilometer level (62 miles). "Okay, but you can't use that number. The Soviets are proposing the hundred kilometer number, and the State Department hasn't agreed to it yet. Give us some maximum sizes and power limits for rockets that can't exceed that altitude." This was done after suitable computer analyses. The resulting exemptions for amateur rockets set forth in 14 CFR 401.5 are based on a rocket that can't exceed 62 statute miles and therefore doesn't penetrate outer space.

Like it or not, 100 kilometers or 62 miles is where outer space begins.

However, between 60,000 feet MSL and 327,360 feet MSL is 267,360 feet of airspace/space over which no one has jurisdiction under current rules, regulations, and treaties.

This will have to be resolved in an international conference within ten years.

Does it matter?

Yes, it does.

Consider the following hypothetical scenario: a spaceship suffers a failure while descending through 62,000 feet MSL for landing. As a result of a sequential catastrophic disassembly of the spaceship as it falls through airspace, damage is caused to people and property on the ground. Which treaty applies?

In an era of expanded and extended space operations with SSTO spaceships, space traffic control will be an absolute necessity for national security just as control of national airspace up to 60,000 feet MSL is today.

A nation launching a space vehicle is required to report the launch and the orbital elements to the United Nations *after the launch takes place* (1974 Convention on Registration of Objects Launched into Outer Space, T.I.A.S. 8480). The North American Air Defense Command (NORAD) routinely detects, tracks, and reports on all orbital and deep space launches. But this occurs after the fact.

This will pose a severe problem for national defense in the future when SSTO spaceships begin operations within and through the atmosphere.

Consider this scenario: a national aerospace defense command detects and tracks a radar target coming over the horizon at 25 times the speed of sound, close to orbital velocity, at very high altitude. That means it's moving at ten miles per second. The people standing watch won't have time to identify it by making telephone calls. They must determine what action to take before the vehicle is over them or past them.

Is the object an incoming ballistic missile warhead or a space weapon dispatched from orbit?

Is it an SSTO returning from orbit? Does it have an announced and pre-filed flight plan in the computer?

Is it the same vehicle that departed on that flight plan?

Is it the daily FedEx flight that left Beijing an hour late and therefore a threat only to UPS?

Does the watch crew alert the ballistic missile defenses? If so, does the officer in charge give the order to shoot or not? If it's a FedEx SSTO, that's one thing. If it's a warhead, that's something else.

If the incoming object isn't transmitting the proper identification code from its radar transponder, rules and procedures must exist that can be implemented at once, no questions asked, and with no blame placed on the defense watch crew for acting in accordance with prearranged procedures.

They will not have time to make a telephone call to get more information. They must know exactly what to do in advance. Rules of engagement must be in place.

The space traffic control is an answer.

The principles of air traffic control must be extended into orbital space with certain modifications unique to the space operations environment.

Space traffic control out to geosynchronous orbit and probably beyond will be required. And soon.

Perhaps, as in Figure 26-1, orbital space from 100 kilometers (62 miles) outward to at least geosynchronous orbit—later to lunar orbit—should be divided into sectors over various continental and oceanic areas around the globe.

Responsibility for detecting, tracking, identifying, and clearing spaceships into and through these sectors would be assigned to multinational or even national organizations in much the same manner that domestic and international air traffic control sectors operate today. If present-day air traffic control technology and procedures are used as a starting point, the development of effective space traffic control shouldn't be difficult. But it will be necessary.

Below 60,000 feet, changes need to be made to existing airspace classifications or new airspace categories be developed.

One must remember that the new spaceships should be treated as a new category of high-performance aircraft that are designed, constructed, tested, certificated, operated, and maintained like air transport airplanes.

How often does an airliner—or even a general aviation airplane—fall out of the sky and crash? When it does happen, it's usually in the middle of a corn field or a forest if it doesn't take place over the 73% of the Earth's surface that's covered by water.

The possibility of spaceships flying over populated areas has caused the Office of Commercial Space Transportation to consider proposed safety regulations. However, in mid-1995 the agency was still operating with the artillery shell paradigm of expendable launch vehicles. Certainly the range safety regulations for expendable rocket shots aren't applicable to true airlinelike commercial space flight. Fortunately, procedures exist for public input and critique of proposed regulations. Full advantage must be taken of these if commercial space advocates are not to be ham-strung by inappropriate regulatory actions.

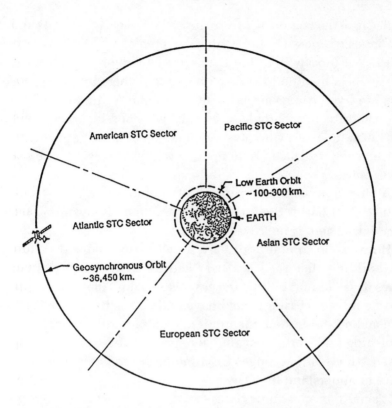

FIGURE 26-1: *Space out to geosynchronous orbit will have to be kept under control and surveillance for safety and security just like the skies of our world. (Drawing by Rick Sternbach and G. Harry Stine.)*

Basically, however, the entire situation of safety certification, space traffic control, airspace management, overflight safety, and the international laws and regulations covering these subjects are confusing, conflicting, internally inconsistent, unworkable, or inapplicable. It's a mess, and people need to get to work now to straighten it out. If it isn't fixed, it may stop commercial space development.

The coming age of commercial space transportation has many facets such as those mentioned here. They've been overshadowed by the intense effort to bring SSTO technology to operational status. The potential is exciting, but it also has many problems left to be attacked and solved.

Because the technology to create SSTO spaceships exists and the economics of operating them are so favorable, we must assume that such spaceships will be built. We must therefore anticipate problems such as the ones discussed here plus others that haven't been thought of yet. It appears prudent to assume technological success and to address these problems as quickly as possible. The problems aren't insoluble. They involve people interacting with people.

However, to make it happen, problems must be treated as opportunities.

A large number of opportunities therefore exist.

In 1977, Charles Sheffield, one of the first space entrepreneurs, observed, "There is no 'royal road' to space."

He went on to point out why we shouldn't depend on anyone else building that road for us: "Unless some other civilization comes to see us and tells us the easy way to do it (and that would be at best a mixed blessing), we'll have to do it the hard way. It will be the usual mixture: some inspiration, a lot of hard engineering, a lot of pushing both for and against by special interest groups, and an overall stimulus that ranges from the desire to make money to the desire to understand the heavens."

It should be clear now that:

a. reusable SSTOs have a long history of development,
b. reusable SSTOs are technically feasible, can be privately built, and can be privately operated economically to produce a profit,
c. space is a place that can be used to make money by doing things of value to others,
d. the sort of space access available with SSTOs will create new markets, some of which are not self-evident,
e. space will not remain the exclusive domain of high-tech gadgetry and people,
f. the federal government cannot do it any more than it could or does run a national airline, much less a suitable passenger railway operation,
g. the legal and regulatory aspects of commercial space transportation haven't been adequately addressed,

h. space transportation is more than SSTO spaceships,

i. we should all be able to go, and

j. people can make money once we do.

In 1965, Phil Bono and Max Hunter started out halfway to nowhere.

In 1993, those who followed in their footsteps got halfway to somewhere.

By the early years of the 21st century, everyone can be halfway to anywhere in the solar system if, instead of a countdown, something like this is heard on a scanner:

"Regional Tower, Astro Seven Delta Charlie, ready for takeoff."

"Astro Seven Delta Charlie, Regional Tower. Cleared for takeoff. Climb unrestricted to orbit on flight plan as cleared. Passing Flight Level Six Zero Zero, contact North American Space Control Center. Have a good flight."

"Thank you, Tower. Full throttle. *Up ship!*"

APPENDIX

Excerpts from a Talk Before the Space Transportation Association

by Daniel S. Goldin, NASA Administrator
February 7, 1996

WE'VE WATCHED THIS GREAT NATION fall to a second-rate power in access to space. Everybody in NASA, DoD, and the aerospace industry ought to hang their heads in shame. I'm as guilty as anyone. We have too much government involvement in the rocket business. We can't go on like this. Things have got to, and must, change. We have a terrible problem because we have to change culture, and, when we change culture, there's risk involved.

I met with the communications industry and asked the lead executives at AT&T, MCI, and others, "What is the biggest thing that we at NASA could help you to do?" They told me, "There is only one thing. We can't continue to compete with ground-based systems if the cost of launching is $10,000 per pound. We need a cost reduction of an order of magnitude, minimum. We want $1,000 a pound or less."

So I made a commitment that this was going to happen. I gave my word to the President of the United States and the Speaker of the House that we're going to make this happen.

We set up the X-33 program where we didn't go along with a traditional government contract and tell the industry what to do. We used a cooperative agreement and put industry in charge. I told the X-33 contractors, "Get with the program and have a little courage. If you don't want the government to tell you what to do, show us some vision and leadership. If you don't want to invest with us, we'll convert to a government-type program, and we'll get to $1000 a pound. It's up to you."

The NASA budget is coming down. It's not pleasant to eliminate jobs and, if we get some further cuts, we'll eliminate some more. I take orders very well, and I'll make it happen.

But the thing that will not happen, the very last light to be turned off, is that commitment to $1000 a pound. We will not back off. We're dead serious. We're not going to do it by throwing money at it. This is not going to be a multi-billion dollar program. We want experimental vehicles that we can fly and crash. We want to design a little and build a little. We're prepared at NASA to cancel some other programs if the budget comes down some more, but we're going to rebuild the launch capability of this nation. We're not stopping at just the X-33. We're going to make some competition. We're going to bring small companies into the rocket business because we'd like the big companies to start acting like little companies.

This is a new NASA, and we're determined to open the space frontier. If industry can do better than NASA, we're going to go to them. If they want to get $1000-a-pound vehicle so they can get rich and open the space frontier, they'll be working with the right people.

I apologize because sometimes I get too intense, but I'm worried about the future of space in this country. You're going to hear this speech again, and again, and again, because, without a space-launch vehicle, there isn't a space program. I didn't come to NASA to watch the shuttle go up and down; I came to NASA to help us open the space frontier, and together we're going to do it.

ACKNOWLEDGMENTS

BECAUSE I KNOW most of the people involved in space transportation—sadly, some of them such as Robert A. Heinlein, Phil Bono, and Daniel O. Graham have passed on—I called upon many friends and colleagues who responded eagerly when I asked for background material. These people include Gary Hudson, Max Hunter, Jerry Pournelle, Bill Gaubatz, the late Daniel O. Graham, Aleta Jackson, Tim Kyger, Henry Vanderbilt, Rick Jurmain, and Paul Hans. I hesitate to mention others because they're deeply involved on career tracks in organizations where their superiors might not appreciate the fact that they shared information with me without getting specific permission to do so. When it was possible for me to do so, I've tried to give them credit in this book for doing what they did, but I wasn't always able to tell the world how they helped me document it. Mind you, no classified, proprietary, or "competition sensitive" information has ever been passed to me by my friends. The careers of brilliant and dedicated people might be jeopardized to the detriment of all of us if I mentioned them at this time. They know who they are, and I will maintain the confidences they've shared with me. Perhaps in a future edition of this book I'll be able to mention their names and document how they helped. But not now. Not while so much is at stake.

BIBLIOGRAPHY

ALDRICH, ARNOLD D., et al, *Access to Space Study Summary Report*, Office of Space Systems Development, NASA Headquarters, Washington DC, January 1994.

BARRY, JIM, *Flying the North Atlantic*, London, B.T. Batsford Ltd., 1987, ISBN 0-7134-5591-8.

BEKEY, IVAN, "Why SSTO Rocket Launch Vehicles Are Now Feasible And Practical; A White Paper," NASA Headquarters, Washington DC, January 4, 1994.

BONO, PHILIP, "ROMBUS—An Integrated Systems Concept for a Reusable Orbital Module—Booster and Utility Shuttle," Douglas Engineering Paper No. 1552, AIAA Preprint No. 63-271, June 16, 1963.

BONO, PHILIP, "The Rocket Sled Launching Technique And Its Implications On Performance of Reusable Ballistic (Orbital/Global) Transport Systems," Douglas Paper No. 4737, presented at the Second International Conference on Planetology and Space Mission Planning, New York Academy of Sciences, New York, NY, October 27, 1967.

BONO, PHILIP, AND GATLAND, KENNETH, *Frontiers of Space*, London, Blandford Press, 1969.

BOORSTIN, DANIEL J., *Hidden History: Exploring Our Secret Past*, New York, Harper & Row, 1987, ISBN 0-679-72223-8.

BROWN, STUART F., "Reusable Rocket Ships: New Low-cost Rides To Space," *Popular Science*, Vol. 244 No. 2, February 1994, pp.49–55.

BYRNE, JAMES, editor, *10:56:20 p.m. EDT, June 20, 1969*, New York, CBS Broadcast Group, 1969 (privately published book about the CBS TV News coverage of the Apollo-11 lunar landing mission).

CALVIN, WILLIAM H., *The Ascent of Mind; Ice Age Climates and the Evolution of Intelligence*, New York, Bantam Books, 1990, ISBN 0-553-35230-X.

CITRON, ROBERT, testimony before the House Committee on Science Subcommittee on Space, Washington DC, March 16, 1995.

CLARKE, ARTHUR C., *Profiles of the Future*, New York, Harper & Row, 1963.

COLE, DANDRIDGE M., *Beyond Tomorrow: The Next Fifty Years In Space*, Amherst, Wisconsin, Amherst Press, 1965.

COLLINS, PATRICK, et al, various papers on space tourism, *The Journal of Space Technology and Science*, Japanese Rocket Society, Vol. 9 No. 1, Spring 1993, Tokyo, ISSN 0911-551X.

COLLINS, PATRICK, et al, "Demand for Space Tourism in America and Japan, and Its Implications for Future Space Activities," *Space Energy & Transportation* (submitted, not yet scheduled for publication).

CORNOG, ROBERT, "Economics of Rocket-Propelled Airplanes," *Aeronautical Engineering Review*, Vol. 15 No. 9, September 1956 and Vol. 15 No. 10, October 1956.

CORREY, LEE, *Shuttle Down*, New York, Ballantine/Del Rey, 1981, ISBN 0-345-29262-6.

DAWSON, TERRY, et al, *Space Launch Oversight Trip*, August 23–September 3, 1993, briefing papers, Washington DC.

DONLAN, THOMAS G., "A Single Stage To Space," *Barron's*, June 21, 1993, p. 10.

DORNBERGER, WALTER R., *V-2*, New York, Viking Press, 1954.

DORNHEIM, MICHAEL A., "NASA Awards RLV Contracts," Aviation Week & Space Technology, June 22, 1994, pp. 22–23.

DULA, ARTHUR M., "Regulation of Private Commercial Space Activities," private paper, ca. 1983, Dula, Shields & Egbert, Suite 6960, Texas Commerce Tower, Houston TX 77002.

DULA, ARTHUR M., "Summary of Present Law and Regulations That Will Effect Non-government Aerospace Transportation Services In the United States," Memorandum to David Hannah, Jr., Space Services., Inc. March 16, 1981, Dula, Shields & Egbert, Suite 6960, Texas Commerce Tower, Houston TX 77002.

GILLIS, JEFF, "Inside the Batcave: the Clementine Mission," *Lunar and Planetary Information Bulletin*, No. 71, May 1994, pp. 2–5.

GLEICK, JAMES, *Chaos; Making a New Science*, New York, Penguin Books, 1987, ISBN 0-14-009250-1.

GOLDIN, DANIEL S., remarks, National Press Club Luncheon, Washington DC, June 20, 1994.

GRAHAM, DANIEL O., *Confessions of a Cold Warrior*, Fairfax, Virginia, Preview Press, 1995, ISBN 0-964-44952-8.

HOESER, STEVEN J., et al, untitled, *The Journal of Practical Applications In Space*, Volume V, Issue No. 3, Spring 1994, pp. 177–184.

HUDSON, GARY C., *Single Stage; The Thirty-year Quest to Develop Real Spaceships*, unfinished draft manuscript, 1994.

HUGHES, BRIAN, quoted in *Space Trans*, Vol. 4, No. 1, January/February 1994, page 6.

HUNTER, MAXWELL W. JR., "Single-Stage Spaceships Should Be Our Goal!" *Nucleonics*, Vol. 21, No. 2, February 1963, pp. 42–45.

HUNTER, MAXWELL W., JR., *Thrust Into Space*, New York, Holt, Rinehart and Winston, 1966.

HUNTER, MAXWELL W., "The SSX SpaceShip, Experimental, Draft II," Space Guild, 3165 La Mesa Drive, San Carlos CA 94070, March 11, 1989.

HUNTER, MAXWELL W., "The SSX—A True Spaceship," *The Journal Of Practical Applications in Space*, Volume 3, Number 4, Summer 1992.

IRVING, CLIVE, *Wide Body: The Triumph of the 747*, New York, William Morrow and Company, Inc., 1993, ISBN 0-688-09902-5.

KAHN, HERMAN, AND BROWN, WILLIAM M., *The Next 200 Years in Space, Final Report,* Croton-on-Hudson, New York, October 23, 1975, HI-2352-RR.

LERNER, PRESTON, "Single Stage To . . . Where?" *Air & Space Smithsonian*, Vol. 8 No. 6, February/March 1994, pp. 44–51.

LOGSDON, JOHN M., *The Decision to Go to the Moon*, Chicago, Illinois, the University of Chicago Press, 1970, ISBN 0-226-49175-7.

MANN, PAUL, "Republicans Seek to Remake NASA," *Aviation Week & Space Technology*, pp. 18–19, December 5, 1994.

MOORMAN, LT. GEN. THOMAS S., JR., "DoD Space Launch Modernization Plan," Briefing to COMSTAC, May 10, 1994.

NAGATOMO, MAKOTO, et al, "Study on Airport Services for Space Tourism," *Sixth International Space Conference of Pacific Basin Societies*, Marina del Rey, California, December 6–8, 1995.

NORRIS, GEOFFREY, "The Short Empire Boats," *Profile Publications Book 4*, Number 84, New York, Doubleday and Company, Inc., 1968.

PETROSKI, HENRY, *The Evolution of Useful Things*, New York, Vintage Books, 1994, ISBN 0-679-74039-2.

PETROSKI, HENRY, *To Engineer Is Human; The Role of Failure in Successful Design*, New York, Vintage Books, 1992, ISBN 0-679-73416-3.

PORT, OTIS, et al, "Is Buck Rogers' Ship Coming In?" *Business Week*, No. 3324, June 21, 1993, pp. 118–120.

POURNELLE, JERRY E., "The SSX Concept," *The Journal of Practical Applications in Space*, Volume 4, Number 3, Spring 1993.

POURNELLE, JERRY E., testimony before the House Committee on Science, Subcommittee on Space and Aeronautics, Washington DC, March 16, 1995.

POURNELLE, JERRY E., et al, *Space and Assured Survival, Report of the Summer 1983 Council Meeting*, Tucson, Arizona, The L5 Society, 1983.

POURNELLE, JERRY E., et al, *America: A Spacefaring Nation, Report of the Fall 1983 Council Meeting*, Tucson, Arizona, The L5 Society, 1983.

POURNELLE, JERRY E., et al, *America: A Spacefaring Nation Again, Report of the Spring 1986 Council Meeting*, Tucson, Arizona, The L5 Society, July 20, 1986.

RYAN, CORNELIUS, editor, *Across the Space Frontier*, New York, Viking Press, 1952 (a collection of articles from *Collier's* magazine, beginning on March 22, 1952).

SHEFFIELD, CHARLES, "Keynote Address," 23rd Annual Meeting, American Astronautical Society, San Francisco, California, October 1977; *Advances in the Astronautical Sciences*, Volume 36, Part 2, "The Industrialization of Space," American Astronautical Society, 1978, ISBN 87703-095-2.

SOCIETY EXPEDITIONS, "Space Tourism Could Drive Space Development," Space Development Conference, Washington DC, April 26–28, 1985.

SPONABLE, JESS M., "Reliable Low Cost Space Transportation," *The Journal of Practical Applications in Space*, Volume 1, Number 4, Summer 1990.

STINE, G. HARRY, *Earth Satellites and the Race for Space Superiority*, New York, Ace Books, 1957.

STINE, G. HARRY, "The Third Industrial Revolution, A Preview of Mankind's Next Cultural Step," *Voyage Beyond Apollo*, U.S.S. Statendam, December 1972.

STINE, G. HARRY, "The Third Industrial Revolution: The Exploitation of the Space Environment," *Spaceflight*, British Interplanetary Society, London, September 1974.

STINE, G. HARRY, *The Third Industrial Revolution*, New York, G. P. Putnam, 1975; New York, Ace Books, 1979, 1981, ISBN 0-441-80665-1.

STINE, G. HARRY, "Government and Industrial Roles In the Initiation of Space Industrialization," 23rd Annual Meeting, American Astronautical Society, San Francisco CA, October 1977; *Advances in the Astronautical Sciences*, Volume 36, Part 2, "The Industrialization of Space," American Astronautical Society, 1978, ISBN 87703-095-2.

STINE, G. HARRY, "Marketing Techniques and Space Industrialization," 23rd Annual Meeting, American Astronautical Society, San Francisco CA, October 1977; *Advances in the Astronautical Sciences*, Volume 36, Part 2, "The Industrialization of Space," American Astronautical Society, 1978, ISBN 87703-095-2.

STINE, G. HARRY, "Wanted: Space DC-3 for Tourists," *Earth/Space News*, 1977.

STINE, G. HARRY, "Economics of Space Industrialization," Joint Symposium, AIAA and World Futures Society, Los Angeles CA, January 1978.

STINE, G. HARRY, "Ticket to Space," *Omni*, March 1979.

STINE, G. HARRY, *The Space Enterprise*, New York, Ace Books, 1980, 1982, ISBN 0-441-77756-2.

STINE, G. HARRY, *Space Power*, New York, Ace Books, 1981, ISBN 0-441-7744-9.

STINE, G. HARRY, *Confrontation In Space*, New York, Prentice Hall, 1981, ISBN 0-13-167437-4.

STINE, G. HARRY, "The Three Dolphin Club," *Analog Science Fiction and Fact*, April 1990.

STINE, G. HARRY, "Space Tourism, the Unbelievable Market, *The Journal of Practical Applications in Space*, Volume 1, Number 4, Summer 1990.

STINE, G. HARRY, *ICBM: The Making of the Weapon That Changed the World*, New York, Orion Books, 1991, ISBN 0-517-56768-7.

STINE, G. HARRY, "Travel and Tourism," *Analog Science Fiction and Fact,* January 1991.

STINE, G. HARRY, "Where Are the Real Engineers?" *Issues in NASA Program and Project Management*, NASA SP-6101 (06), National Aeronautics and Space Administration, Washington DC, Summer 1993.

STINE, G. HARRY, "The SSTO Operational Environment," *The Journal of Practical Applications in Space*, Volume 4, Number 2, Winter 1993.

STINE, G. HARRY, "The Rooster Crows at White Sands," *Analog Science Fiction and Fact*, Vol. CXIV No. 6, May 1994.

STINE, G. HARRY, "Why Build Experimental Vehicles?" guest editorial, *Analog Science Fiction and Fact*, Vol. CXV No. 7, June 1995.

STINE, G. HARRY, "The Fractional Orbital Transportation System," *The Journal of Practical Applications in Space*, Volume 6, Number 1, Fall 1994.

STINE, G. HARRY, "Some Legal and Regulatory Issues of the Commercial Space Age," *Space Energy & Transportation*, Volume 1, Number 2, Spring 1996.

STINE, G. HARRY, AND HANS, PAUL C., "Economics of Hypersonic Vehicles and Spaceplanes," USAF Hypersonic Conference, Arlington VA, July 1990.

STINE, G. HARRY, AND HANS, PAUL C., "Economic Considerations of Hypersonic Vehicles and Spaceplanes," AIAA-90-5267, AIAA Second International Aerospaceplanes Conference, Orlando FL, 29–31 October 1990.

STINE, G. HARRY, AND SMITH, W.C., "Laughing All the Way to Orbit," *Analog Science Fiction and Fact*, February 1988.

STINE, G. HARRY, et al, "Financing Alternatives for Space Industrialization," 4th Princeton Conference on Space Manufacturing Facilities, AIAA and Princeton University, Princeton NJ, 1979.

SUTTON, GEORGE P., *Rocket Propulsion Elements, Third Edition*, New York, John Wiley & Sons, 1963.

ZUBRIN, ROMERT M. AND CLAPP, MITCHELL BURNSIDE, "Black Horse: One Stop to Orbit",*Analog Science Fiction and Fact*, Vol. CXV No. 7, June 1995.

_____ , "Treaty on Principles Governing the Activities of States in the Exploration an Use of Outer Space," Article 7, United Nations, New York, 1967.

_____ , *Final Report On Project, Single-Stage Earth-Orbital Reusable Vehicle, Space Shuttle Feasibility Study*, DRD MA-095-U4, TR-AP-71-4, contract NAS8-26341, June 30, 1971. Chrysler Corporation Space Division, P.O. Box 29200, New Orleans LA.

_____ , Hearings before the Committee on Science and Technology, U.S. House of Representatives, Ninety-fifth Congress, Second Session, January 24–26, 1978, U. S. Government Printing Office.

_____ , *Satellite Power System Concept Development and Evaluation Program, Reference System Report*, US Department of Energy, Office of Energy Research, Washington DC, DOE/ER-0023, October 1978.

_____ , *Satellite Power System Concept Development and Evaluation Program, Preliminary Assessment*, US Department of Energy, Office of Energy Research, Washington DC, DOE/ER-0041, September 1979.

_____ , *Space Commerce: An Industry Assessment*, U.S. Department of Commerce, Washington DC, May 1988.

_____ , "Air Cargo, 1988 Traffic and Assessment," *Distribution* Magazine, February 1989.

_____ , *James Gleick's Chaos: The Software*, Autodesk, Inc., Bothell, Washington, 1990.

_____ , *Report To Congress, Single Stage Rocket Technology,* Ballistic Missile Defense Organization, Department of Defense, Washington DC, 1992.

_____ , "Key Staffer Eyes Launch Czar for U.S.," Space Business News, Vol. 11, No. 24, December 7, 1993, pp. 1–2.

_____ , *National Space Transportation Policy,* Office of Science and Technology Policy, The White House, Washington DC, August 5, 1994.

_____ , *Commercial Space Transport Study Final Report*, jointly published by Boeing, General Dynamics, Lockheed, Martin Marietta, McDonnell Douglas, and Rockwell, May 1994. (This report is unattributed; the author obtained his copy from Dana Andrews, Mail Stop 8C-09, Boeing, P.O. Box 3707, Seattle WA 98124.)

_____ , Hearings before the House Committee on Science and Technology, Washington DC, July 19, 1994.

_____ , National Space Transportation Round Table on Reusable Launch Systems, National Space Society, Washington DC, September 14, 1994 (unedited transcript and videotape).

_____ , *A Cooperative Agreement Notice, Reusable Launch Vehicle (RLV) Advanced Technology Demonstrator X-33,* CAN 8-1, Program Development Directorate/PAO1, NASA George C. Marshall Space Flight Center, Huntsville AL, January 12, 1995. (Draft dated 19 October 1994.)

_____ , NFPA 50A *Liquefied Hydrogen Systems*, Quincy, Massachusetts, National Fire Protection Association.

_____ , Code of Federal Regulations, 14 CFR Part 21 through 14 CFR Part 29, Federal Aviation Regulations, U.S. Government Printing Office, Washington DC.

_____ , Code of Federal Regulations, 14 CFR Part 400 to 415, Office of Commercial Space Transportation, Department of Transportation, OCST Regulations, U.S. Government Printing Office, Washington DC.

_____ , "X-33, X-34 Contractors Selected for Negotiations," NASA Press Release 95-23, NASA Headquarters, Washington DC, March 8, 1995.

ADDRESSES

The Journal of Practical Applications in Space and (after January 1996) *Space Energy & Transportation*, Aleta Jackson, Editor, a publication of the Space Transportation Association, 2800 Shirlington Road, Suite 405A, Arlington, VA 22206, telephone 703-671-4111, e-mail highfrontier@bix.com.

National Space Society, 922 Pennsylvania Avenue S.E., Washington, DC 20003, telephone 800-543-1280.

The Space Access Society, Henry Vanderbilt, Executive Director, 4855 E. Warner Road #24-150, Phoenix, AZ 85044, telephone 602-431-9283, e-mail hvanderbilt@bix.com.

The Space Frontier Foundation, New York, NY, telephone 212-387-7887, e-mail openfrontier@delphi.com.

INDEX

ABOUT THE AUTHOR

G. HARRY STINE is uniquely qualified to write this book. He began his professional career in 1951 as an engineer at the White Sands missile testing range in New Mexico and has served as a consultant to NASA and several aerospace companies. A founder of the Citizen's Advisory Council on National Space Policy, he is presently one of the thirteen members of the Arizona Space Commission. He is a former member of the NASA Technology and Commercialization Advisory Committee.

He is the author of more than twenty science-fact books. Among them is *The Third Industrial Revolution*, the first book to reveal what we can do in space to make money. Another, *Confrontation In Space*, addressed the basic military realities of space and was widely read at the service academies and military/naval staff colleges. His *Handbook for Space Colonists* became a standard reference work about human beings in space. He also wrote the first book about the space shuttle, *Shuttle Into Space*.

Halfway To Anywhere is a unique insider's view of the unheralded revolutionary changes taking place in space transportation today, changes that will affect the lives of everyone within a decade and allow anyone to go cheaply and easily into this new frontier.